T0248764

Encyclopedia of Alternative and Renewable Energy: Biomass Utilization and Bioreactors

Volume 09

Encyclopedia of Alternative and Renewable Energy: Biomass Utilization and Bioreactors

Volume 09

Edited by **Brad Hill and**
David McCartney

New York

Published by Callisto Reference,
106 Park Avenue, Suite 200,
New York, NY 10016, USA
www.callistoreference.com

Encyclopedia of Alternative and Renewable Energy:
Biomass Utilization and Bioreactors
Volume 09
Edited by Brad Hill and David McCartney

International Standard Book Number: 978-1-63239-183-4 (Hardback)

Printed in the United States of America.

Contents

Preface

This book has been an outcome of determined endeavour from a group of educationists in the field. The primary objective was to involve a broad spectrum of professionals from diverse cultural background involved in the field for developing new researches. The book not only targets students but also scholars pursuing higher research for further enhancement of the theoretical and practical applications of the subject.

Research in the field of biomass and related disciplines has increased with time. Awareness regarding pros and cons of biomass, sustainable use of resources and biomass sources has brought in the concept of biorefineries. This is evident from growth in biomass research and increasing attention towards biofuels. This book covers various disciplines of biomass and bio-reactors dealing with various aspects of bio-digestion and overall bio-reactor processes.

It was an honour to edit such a profound book and also a challenging task to compile and examine all the relevant data for accuracy and originality. I wish to acknowledge the efforts of the contributors for submitting such brilliant and diverse chapters in the field and for endlessly working for the completion of the book. Last, but not the least; I thank my family for being a constant source of support in all my research endeavours.

Editor

Bio-Reactors

Biomass Digestion to Produce Organic Fertilizers: A Case-Study on Digested Livestock Manure

Alessandra Trinchera, Carlos Mario Rivera, Andrea Marcucci and Elvira Rea

Additional information is available at the end of the chapter

1. Introduction

Biogas production by anaerobic digestion of organic wastes coming from agricultural practices is one of most promising approach to generate renewable energy, giving as end-product a digested organic biomass with specific characteristics useful for soil fertilization. This last aspect represents an opportunity in relation to the need to close the nutrient cycles within the agricultural and natural ecosystems, particularly in specific systems underwent to a constant resources depletion, as those of Mediterranean area, where the C-sink loss represents one of the main causes of desertification [1], [2]. The composting process was yet identified as one of the promising answers to the need of soil organic matter conservation, such as the addition to the soil of different organic materials of different origins [3], but the anaerobic digestion could represent an effective further step able to guarantee the recycle of nutrients, coupled with an environmental-friendly energy production [4].

Particularly, anaerobic digestion of livestock manures allows us to achieve several purposes: i) renewable energy generation; ii) reduction of nitrate leaching in livestock exploitations, iii) production of an organic biomass as by-product employable as organic fertilizer [5]. Actually, digestates coming from livestock manure give biomasses characterized by biologically stable organic matter and relevant nitrogen content; these traits suppose that these biomasses may be usefully utilized as N-fertilizers and soil organic amendments in agriculture, but also as a component of growing media in pot horticultural cultivation.

It should be remarked that production of greenhouse horticultural crops requires the use of growing media of high quality, with specific physical-chemical characteristics. Being the peat, organic component traditionally used in substrates formulation, a non-renewable resource (its extraction involving many environmental issues), it becomes ever more urgent the need to individuate alternative organic materials with the same functioning, such as composts or products coming from biogestion processes [6]. Nevertheless, the use of biodigestate as amendment, still now not allowed in Italy, should provide the declaration of stability parameters for organic matter, since the utilization of matrices not properly stabilized could lead to the risk of high fermentescibility of organic components and, thus, consequent phytotoxicity phenomena [6],[7]. The stabilization of organic matter actually involves the mineralization of the most labile organic fraction, with the following decrease of C/N ratio; this means physical, chemical and biological changes of the starting material and, thus, the decrease of porosity, increase of pH, CEC, bulk density and salinity, due to a concentration of organic compounds which, generally, are characterized by lower molecular weight respect to the starting ones, more resistant to microbial degradation [8].

The amendment properties of composts and biodigestates could be assessed by different analytical methods, such as the isoelectrophoretic techniques (IEF) [9],[10],[11]. Results obtained after IEF characterization of the extractable fraction in alkaline environment of dried vine vinasse (an anaerobically digested solid residues, constituted by exhausted stems, skins and grape seeds, obtained after distillation of the "grappa", the Italian "acquavite"), showed an increase of the extractable organic components in alkaline environment, with a higher content in less acidic organic fraction, probably due to a "concentration effect" of the more complex organic components not, or only partially, degraded during the anaerobic digestion process.

Other works demonstrated also that the same dried vine vinasse, applied together with other mineral components in growing media, was able to increase nutrient availability [12] and express a sort of biostimulant activity on plant roots [13]. Study performed by optical microscopy demonstrated that digested vine vinasse in combination with clinoptilolite addition, promoted maize roots development, by increasing mucigel production by root tip and thus favoring the following solubilization and uptake of nutrients by plant from the added organic biomass (Figure 1).

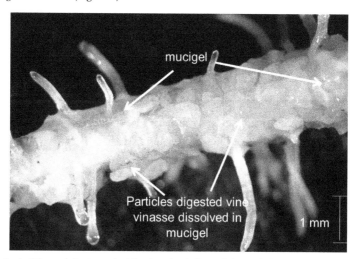

Figure 1. Root of Zea mais L., treated with micronized clinoptilolite and digested vine vinasse.

In relation to the N-fertilizer attitude of a biodigestate, it has a particular relevance in the case of the biodigestion of animal manure. The anaerobic process allows to maintain constant the total N amount of the original material, even if the organic N is mainly transformed into ammonia, the mineral form most easily available to the crops. The separation between the liquid and the solid fractions after biodigestion allows to recover the ammonium-N in the fluid fraction and the residual organic matter in the solid fraction, so to emphasize the different characteristics of the two fractions: the first one, usable as a typical N-fertilizers, able to furnish nutrient supply to plant; the second one as organic amendment, able to supply organic matter to soil and then improve its chemical, physical and biological characteristics. A proper composting process applied to the solid fraction of this digestate could lead to a further improvement of the biomass, by promoting its biochemical stability and giving those amendment properties yet described, which constitute the adding value of the final product.

What is relevant is that the risk of nitrate leaching in water represents the main limitation to the direct application of not pre-treated livestock manure to soil: effectively, its amendment properties, linked to soil organic matter addition, often go in conflict with the Council Directive 91/676/EEC on the "Water protection from nitrates" [14], especially in vulnerable zones, such as those of Italian northern regions. In this sense, the biodigestion process represents a great opportunity to utilize livestock manure digestates as N fertilizer, potentially allowing to overcome the limit of 170 Kg N ha^{-1} year^{-1} superimposed by EU Nitrates Directive, since the high N-efficiency coefficient of these digestates, similar to that of soluble mineral fertilizers [15]. The solid fraction, containing organic matter already partially stabilized, could also permit its application during the winter season, provided the obtained amendment has a constant composition and a fraction of slow release-N, eventually increased by a following composting process [16],[10],[17]. On the other hand, the anaerobic conditions ensure the formation of high amount of ammonium during the organic matter degradation process, without incurring in the subsequent oxidation into nitrate [18],[4]. Being the ammonium the N-form more rapidly assimilated by the crops, this could be a further element in favor of the utilization of biodigestates as components of growing media, effectively conjugating the physical amendment properties with those chemical, connected with fertilization.

In 2007, the Italian ministry of agriculture financed a Research Project on the "Anaerobic digestion of livestock manure and EU Nitrates Directive – Effect due to the anaerobic digestion on N availability in livestock manure for overcoming the limit of 170 Kg N ha^{-1} year^{-1} superimposed by imposto EU Nitrates Directive". The main aim of the reported study was to verify the N-fertilizer properties of a digestate coming from a swine livestock manure, taking into account the possibility to utilize this processed agricultural waste of animal origin for limiting the risk of environmental pollution. Hereafter, results related to the effect of the digested and not-digested solid fraction of this bovine livestock manure applied as N-organic fertilizers on lettuce growth in a greenhouse experiment are reported.

2. Materials and methods

2.1. Soils characterization

Two different soils of Northern Italy were chosen in a specific area located in Pedemont Region (Vercelli, Italy) characterized by cold and humid climate, in relation to their recognized vulnerability for nitrate leaching. The two soils, Tetto Frati (A) and Poirino (B), sampled at 0-30 cm, were characterized for main chemical, physical and biochemical parameters: pH, CEC (meq 100 g^{-1}), total organic C % [19] and total N (mg Kg^{-1}) [20], texture (sand %, loam %, clay %), bulk density (g cm^{-3}), basal respiration [(mg$_{C-CO2}$ ×Kg$_{soil}$$^{-1}$)×day^{-1}], C microbial biomass content (mg kg^{-1}) [21], metabolic quotient qCO$_2$ [(mg$_{C-CO2}$×mg C$_{biom}$$^{-1}$×h)×10^{-3}], mineralization quotient qM, [22] and C microbial biomass/C organic ratio (%) [23].

2.2. Biomasses characterization

The organic biomasses, not-processed and processed, were supplied by the anaerobic digestion plant of the "Research Centre for Animal Production - Foundation Centre for Studies and Research" (C.R.P.A. – F.C.S.R.   S.p.A., Reggio Emilia, Italy). Their main chemical characteristics of solid fractions of swine livestock manure are below reported:

pH: not-digested = 7.0   digested = 8.0

N content (as received):   not-digested = 0.47%   digested = 0.39%

N content (dry matter)   not-digested = 2.9%   digested = 3.0%

C organic (dry matter):   not-digested = 43,5%   digested = 35,9%

In order to evaluate the organic matter stability of the two biomasses, both the digested and not-digested solid fractions of swine manure were previously analyzed by isoelectrofocusing (IEF) technique. Organic matter extraction was carried out on 2 g of each biomass with 100 mL of a solution NaOH/Na$_4$P$_2$O$_7$ 0.1N for 48 hours at 65°C. After centrifugation and filtration, the extracts were stored at 4°C under nitrogen atmosphere. Ten millilitres of NaOH/Na$_4$P$_2$O$_7$ extracts were dialysed in 6,000-8,000 Dalton membranes, lyophilised and then electrofocused in a pH range 3.5-8.0, on a polyacrylamide (acrylamide/bis-acrylamide: 37.5/1) slab gel, using 1 mL of a mixture of carrier ampholytes (Pharmacia Biotech) constituted by: 25 units of Ampholine pH 3.5-5.0; 10 units of Ampholine pH 5.0-7.0; 5 units of Ampholine pH 6.0-8.0. A prerun (2h 30'; 1200V; 21mA; 8W; 1°C) was performed and the pH gradient formed in the slab gel was checked by a specific surface electrode. The electrophoretic run (2h; 1200V; 21mA; 8W; 1°C) was carried out loading the same C amount of water-resolubilised extracts (1 mg C×50 L^{-1} × sample^{-1}). The bands obtained were stained with an aqueous solution of Basic Blue 3 (30%) for 18 h and then scanned by an Ultrascan-XL Densitometer (Amersham – Pharmacia) [10].

2.3. Pot trial on lettuce (experimental plan, plant biometric survey and elemental analysis)

In a 300 m² greenhouse, lettuce (Lactuca sativa L., var. Romana) were transplanted into 2 L and 16 cm diameters pots, containing the A and B soils; growing density was 16 pots m². The experiment was performed from April to June, 2009, at temperatures range of 15-28°C.

Treatments consisted in a factorial combination of two increasing N doses (200 and 400 $kg_N \times ha^{-1}$), applied as solid fraction of digested swine livestock manure, not-digested solid fraction of swine livestock manure and granular urea [$CO(NH_2)_2$], taken as conventional mineral fertilization; not-fertilized soils were also considered as control treatments. Even if the N recommended dose for lettuce growth is about 90-100 $kg_N \times ha^{-1}$, the choice of such high N supply was made on the basis of the need to overcome the limit of 170 Kg_N ha^{-1} by substituting these organic biomasses rich in N to the mineral fertilization, without incurring in undesired effects on plant and environment: the dose of 400 $kg_N \times ha^{-1}$ was just applied for evaluating its potential phyto-toxicity on lettuce in relation to the different fertilization treatments. The corresponding fertilizers'doses per pot were: 128.6 g and 257.2 g for ND, 153.2 g and 306.3 g for D and 1.4 g and 2.6 g for urea.

Treatments were arranged in a randomized complete-block design with six replicates. Drip irrigation was managed in relation to plant water-demand, as reported in Figure 2.

Figure 2. Example of pot cultivation of lettuce in greenhouse; drip irrigation was used for guaranteeing daily water supply to the plants .

Lettuce plants were harvested 70 days from sown; biomass dry weight (g), dry matter (%), leaf area (cm^2) and leaf number were determined for each plants. In order to evaluate the effect of alternative fertilization treatments on micro and micronutrient uptake by lettuce, N, P, K, Mg, Cu, B, Fe, Mn leaf contents, plant material was incinerated at 400°C for 24 hours; ashes were then redissolved in HCl 0.1N and the supernatant filtrated to obtain a limpid solution; the nutrient content was determined by simultaneous plasma emission spectrophotometer (ICP-OES) on obtained solution and calculated in relation to dry matter.

2.4. N use efficiency

After analysis of N leaf tissue content (%) by Kijeldhal method, N-use efficiency (NUE %) was calculated, as the percentage of the N uptaken by the lettuce plant respect to N supplied by the fertilizer.

In order to study the long-term effect of the alternative fertilization approaches, the soil residual N at the end of experiment was obtained after Kjeldhal digestion and titrimetric determination [20]. Then, the available N-NO$_3$ and exchangeable N-NH$_4$ in the soils were determined after extraction of 4 g of each soil in 40 mL of KCl 0.2 N solution and subsequent colorimetric analysis of the supernatant by Automatic Analyzer Technicon II.

2.5. Statistical analysis

Plant biometric and soil N data were evaluated by ANOVA to verify the statistical differences of the tested parameters in relation to the different fertilization treatments.

Elemental data were analysed using vector analysis, which allows the simultaneous evaluation of plant dry weight and nutrients content in an integrated graphic format [24],[25]. Elemental data in relation to the different treatments in soils A and B were normalized with respect to urea al 200 kg$_N$ ha^{-1}, taken as reference treatment.

3. Results and discussion

3.1. Soils characterization

In Table 1, the chemical-physical and biochemical parameters of the two considered A and B soils are reported.

The comparison between soils characteristics showed that A was a typical sandy-loam soil, with a lower organic C, lower C microbial biomass content, lower C microbial biomass/C organic ratio and higher mineralization quotient respect to B, which was a loamy soil, with higher organic C, higher C microbial biomass content, higher C microbial biomass/C organic ratio and lower mineralization quotient (Figure 1). On the basis of these results, the two soils were defined as a low biological fertility soil (Soil A) and a medium biological fertility soil (Soil B) [[23],[26].

The choice of these two different soils was performed in order to evaluate the effect played by the different soil characteristics on the behaviour of the digested and not-digested materials utilized in the experiment: it is well known that in all biologically mediated processes, such as the mineralization/immobilization of N coming from organic fertilizers, the microbial biomass has a key-role in addressing the degradation of organic substrate and making available N to the plant [27],[28],[29].

On the basis of what above discussed, the results were elaborated by considering the two soils as two independent experimental units, where the same experimental design was applied.

Parameter	Soil A	Soil B
Sand (%)	48,4	15,8
Loam (%)	43,1	75,6
Clay (%)	8,5	8,6
pH	8,2	6,1
C organic (g kg^{-1})	12	17
N total (mg kg^{-1})	0,83	0,81
CEC (meq 100g^{-1})	8,2	12,5
Bulk density (g cm^{-3})	1,34	1,45
Basal respiration (mg $_{C-CO2}$ Kg$_{soil}^{-1}$) day^{-1}	7,6	5,7
C microbial biomass (mg kg^{-1})	172	281
C microbial biomass/C organic (%)	1,43	1,65
qCO$_2$ (mg $_{C-CO2}$ mg $_{Cbiom}^{-1}$ h) 10^{-3}	1,85	0,85
qM (%)	4,3	3

Table 1. Main chemical, physical and biochemical parameters of A and B soils.

3.2. Biomasses characterization

In Figure 3, isoelectric focusing of the extracted organic matter from not-digested and digested solid fraction of swine manure utilized in the experimental activity is reported.

As reported in literature [30],[31], the more humified is the organic matter, the less acidic and higher in molecular weight are the organic compounds which compose it. This information, which is valid for soil humic matter but also for humic-like compounds in biomasses, allows to use IEF technique in order to separate organic matter extracted from different materials. Taking into account that, generally, organic matter focused in a pH range between 3.0 and 5.5, obtained results indicated that the organic matter from digested solid fraction was better stabilized respect to the not-digested one: in fact, the increasing peaks at pH >4.7 indicated a good stabilization of extracted organic matter from digested material, that means increased amendment properties of this material [16].

Effectively, the increase of the peaks' area in the pH region higher than 4.5 indicates firstly, that during the biodigestion changes in chemical composition of the organic material took place and, secondly, that these transformations led to the constitution of a final biomass made by less acidic compounds, higher average molecular weight molecules, that means more chemically stable material [10].

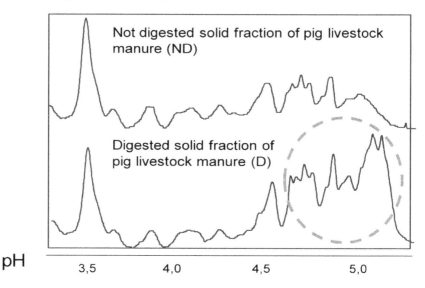

Figure 3. Isoelectrofocusing of extracted organic matter in alkaline environment from not-digested and digested solid fraction of pig livestock manure.

3.3. Pot trial on lettuce (plant biometric survey and elemental analysis)

In Figure 4, lettuce plants differently fertilized in the two soils at the end of the experiment (60 days) are shown.

The effect of the different treatments and soils is evident when comparing the lettuce growth. The phytotoxic effect of urea at highest dose was dramatically shown in Soil A, while at the same urea dose, in soil B the lettuce was able to increase its development although the excess of N mineral supply: this result attested clearly the role of soil in influencing the actual availability of nutrients, and in particular of N, for plant uptake. It was promising the good performances of both the not-digested and the digested solid fraction of livestock manure at the highest N dose, particularly in Soil A: even if the 400 kg$_{N}$×ha^{-1} supplied by urea gave the worst result on lettuce, apparently the excess of N added with the organic biomasses did not determine any decrease of lettuce growth, but on the contrary, a very good development of lettuce foliage (Figure 5).

This aspect is a positive point in order to propose the increase of the limit rate of 170 kg$_{N}$×ha^{-1} year^{-1} for digestate application to soil, especially because these results were obtained in a soil with a sandy texture, low organic C content and, consequently, particularly vulnerable for nitrates.

(ND = not-digested solid fraction of livestock manure; D = digested solid fraction of livestock manure; 200 = 200 kg$_N \times$ha^{-1}; 400 = 400 kg$_N \times$ha^{-1}).

Figure 4. Lettuce grown in relation to the different fertilization treatments in soil A and B.

(400 = 400 kg$_N$×ha^{-1}).

Figure 5. Example of lettuce leaves development in relation to the different fertilization treatments in Soil A.

The quantitative results related to biomass production and quality of plants are reported in Table 2.

Soil A	Dry weight (g plant^{-1})	Dry matter (%)	Total leaf area (cm^2)	Number of leaves
Not fertilized	1,4 b	5,2 b	450 b	15 a
Urea 200 Kg ha^{-1}	2,9 d	4,8 b	1450 d	24 c
Urea 400 Kg ha^{-1}	0,4 a	5,4 c	180 a	13 a
Not digested 200 Kg ha^{-1}	1,2 b	4,4 a	730 c	17 b
Not digested 400 Kg ha^{-1}	2,5 c	7,3 d	750 c	18 b
Digested 200 Kg ha^{-1}	1,9 c	5,0 b	760 c	17 b
Digested 400 Kg ha^{-1}	2,4 c	5,0 b	850 c	18 b
Soil B	**Dry weight (g plant^{-1})**	**Dry matter (%)**	**Total leaf area (cm^2)**	**Number of leaves**
Not fertilized	0,7 a	1,9 a	900 b	18 ab
Urea 200 Kg ha^{-1}	3,5 c	3,7 c	2000 c	23 c
Urea 400 Kg ha^{-1}	0,8 a	5,2 cd	1300 bc	22 c
Not digested 200 Kg ha^{-1}	1,7 b	3,7 c	850 b	18 ab
Not digested 400 Kg ha^{-1}	1,6 b	3,2 b	1250 bc	20 b
Digested 200 Kg ha^{-1}	0,7 a	2,1 a	700 a	17 a
Digested 400 Kg ha^{-1}	1,6 b	3,5 c	1000 b	20 b

Table 2. Lettuce dry weight, dry matter, total leaf area and number of leaves obtained after fertilization treatments in A and B soils (average value; different letters means significant differences at P-level<0.05).

Firstly, again a strong "soil effect" was recorded in relation to plant growth parameters for all the treatments, due to the different chemical-physical characteristics and biological fertility of the two soils. Not fertilized plants showed a limited vegetative development while, as expected, lowest urea dose (200 kg$_N$×ha^{-1}) gave the best plant growth (6.7 g plant^{-1}), as confirmed by recorded parameters as shoot dry weight, percentage of dry matter, leaf area and leaf number, especially in B Soil. On the contrary, urea at 400 kg$_N$×ha^{-1} dramatically depressed plant growth in Soil A (0.8 g plant^{-1}), due to evident toxicity phenomena.

Digested and not digested biomasses gave best results when applied at the higher dose respect to the lowest one; actually, in treatments with both the biomasses at 400 kg$_N$×ha^{-1}, plant parameters were closer to those obtained with urea at 200 kg$_N$×ha^{-1}. It is relevant that in Soil A the application of both the digested and the not-digested solid fractions of livestock manure at 400 kg$_N$×ha^{-1} gave weight parameters higher than those observed at 200 kg$_N$×ha^{-1}. Otherwise, in Soil B, only fertilization with digested solid fraction of livestock manure at 400 kg$_N$×ha^{-1} gave an increase of all the tested parameters respect to the 200 kg$_N$×ha^{-1} dose, while the not digested biomass did not show any differences among the N rates.

No toxicity phenomena were detected also at the highest doses of added biomasses and this is an important and positive result in the scope of utilizing these materials in substitution of mineral fertilizer also with high doses of N supply without collateral effect on plant.

For better clarify the effect of the alternative fertilization treatments on lettuce plant parameters in relation of the two soils, radar graphs are reported in Figure 6.

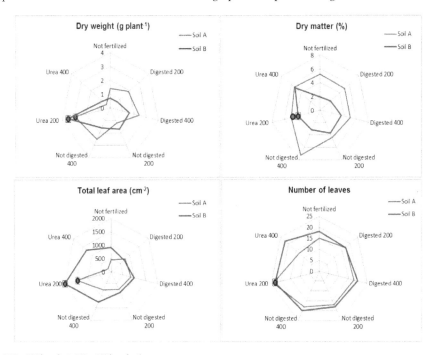

(200 = 200 kg$_N$×ha^{-1}; 400 = 400 kg$_N$×ha^{-1}).

Figure 6. Radar graphs of plant parameters related to the fertilization treatments in soil A and B.

Lettuce number of leaves was little affected by soil characteristics, while dry weight, dry matter and total leaf area were evidently influenced by both the soil and the fertilization. On general terms, in the Soil A the percentages of dry matter of lettuce were higher respect to the corresponding values recorded in Soil B; on the contrary, the total leaf areas were lower in Soil A than in Soil B. The water uptake seems to have had a great importance in the two considered systems: probably, it was strongly affected by physical characteristics of the two soils. In Soil A, with about 50% of sand content, the water was less available to plant because of its lower water retention capacity respect to Soil B, so to determine the tendency to reduce the lettuce total leaf area and increase leaves dry matter. On the contrary, in the loamy Soil B, the higher water availability determined a decrease in percentages of lettuce dry matter and the increase of leaf areas, so to give an indication of the dependence of lettuce quality mainly from soil characteristics rather than the fertilization treatments. Anyway, taking into account 200 $kg_N \times ha^{-1}$ of urea as the reference dose for lettuce production, it is relevant that the not-digested solid fraction of livestock manure at 400 $kg_N \times ha^{-1}$ gave the highest value of lettuce dry matter among all the treatments.

For evaluating macro (N, P, K, Mg) and micronutrients (Cu, B, Fe, Mn) use efficiency, biomass dry weight and elemental concentrations were plotted, including curved content isoclines [32],[33]. Each point on the bidimensional plot represent a vectors, taken as control equal to 100% both the concentration and the related dry weight obtained after addition of 200 $kg_N \times ha^{-1}$ urea (intersection point) in graphs (Figure 7 and Figure 8).

Plant tissue composition was significantly affected by the different treatments, depending on the added materials and rate. In relation to both the macro and the micronutrients, it should be remarked that all results were shifted to the limiting vector space (left side of the plot, under 100% lettuce dry matter, corresponding to urea at 200 $kg_N \times ha^{-1}$), representing the reducing growth treatments.

In relation to N (Figure 7), the phytotoxic effect of mineral fertilizer at 400 $kg_N \times ha^{-1}$ is particularly evident in Soil A, where the increase of concentration of N corresponded to the greatest decrease of lettuce dry matter respect to the control; the same severe phytotoxicity was not recovered after treatments with organic biomasses. The most promising results were obtained after addition of not-digested and digested solid fraction of livestock manure at highest dose, giving a dry matter similar to those obtained with the control mineral fertilization, but with a net decrease in N uptake: this finding attests that the N use efficiency was particularly high when 400 $kg_N \times ha^{-1}$ of both organic materials were added to Soil A, since the lack in N prompt availability was not so heavy in limiting plant growth. Such a positive result was not so evident in Soil B, because of the clear reduction of lettuce dry matter (about -50%) after treatments with digested and not-digested materials respect to urea at 200 $kg_N \times ha^{-1}$, even if the 400 $kg_N \times ha^{-1}$ urea application gave a tendency to an excess of N consumption by lettuce, which did not correspond to an increase of dry matter production.

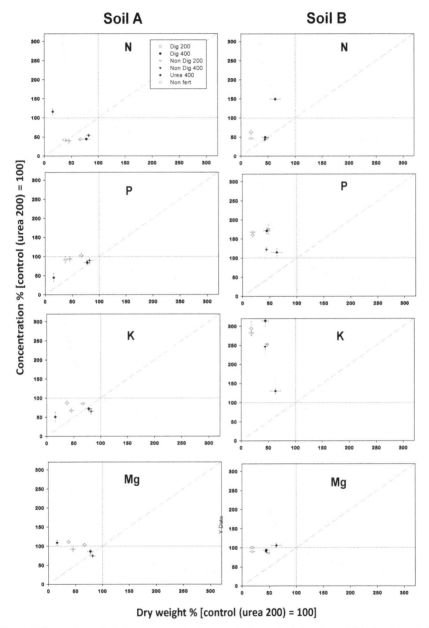

Figure 7. Comparison of relative macronutrients concentration and relative dry weight of aerial part of lettuce plant at the end of growing cycle. Control values of dry weight and concentration used as reference (point of isolines intersection).

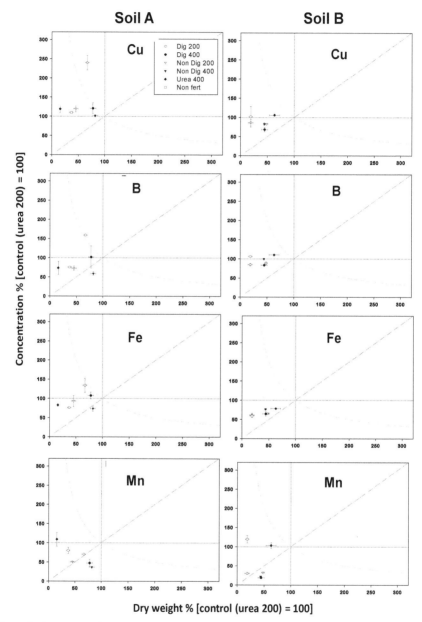

Figure 8. Comparison of relative micronutrients concentration and relative dry weight of aerial part of lettuce plant at the end of growing cycle. Control values of dry weight and concentration used as reference (point of isolines intersection).

Similar results were obtained also for Mg (Figure 7), since again this nutrient concentration seemed to be another limiting factor for lettuce growth, in both the soils.

Different behaviour was recorded for P and K (Figure 7): their related uptakes were strongly affected by both the soils characteristics and the fertilization treatments: while in Soil A the limiting factor for lettuce growth was clearly the soil P and K availability (left space of the graph, under 100% of nutrient concentration for the urea control), in Soil B the effect was opposite (especially for K), since the excess of nutrients appeared to be the main cause of plant growth decrease (left space of the graph, above 100% of nutrient concentration for control 200 $kg_N \times ha^{-1}$), determining a typically defined "nutrient luxury consumption".

In relation to micronutrients Cu, B, Fe and Mn (Figure 8), sometimes their deficiency represented the main limiting factor for lettuce growth (as Fe in Soil B for all the treatments), sometimes the excess of their concentration could have again determined a luxury consumption (as Cu and B in Soil A, after addition of 200 $kg_N \times ha^{-1}$ of digested livestock manure). It is interesting the effect played by soils on Mn lettuce uptake: in Soil A, after addition of 200 $kg_N \times ha^{-1}$ of digested livestock manure, both the Mn uptake and the lettuce dry weight were reduced only of about 30% respect to the values posed to 100% for urea control; on the contrary, in Soil B, the same parameters strongly decreased of about 80%, so confirming the role of soil chemical and physical characteristics on nutrient availability in relation to the different treatments.

3.4. N use efficiency

In Table 3, N use efficiency (calculated as percentage of $N_{uptaken}/N_{fert}$), the residual soil N, the available soil N-NO$_3$ and N-NH$_4$ in relation to the different fertilization treatments in soil A and B are reported.

As far as the lettuce N use efficiency is concerned, in both the soils the highest NUE % was obtained after urea application, with the exception of urea at the rate of 400 $kg_N \times ha^{-1}$ in Soil A, which caused an evident phytotoxicity, already manifested by the reduction of lettuce plants growth. In the same soil, higher NUEs were obtained after application of the digested and not-digested solid fraction of livestock manure at the higher 400 $kg_N \times ha^{-1}$ rate. In B soil, while application of the digested biomass at both the doses gave similar NUE values, a quite doubled NUE was obtained after addition of not digested livestock manure at 200 $kg_N \times ha^{-1}$: these positive results could be explained by the presence of higher amount of promptly available N in the not-processed livestock manure respect to the digested one. Anyway, the recorded N use efficiency obtained after the application of both the biomasses was, as expected, lower respect to the urea application at the same rates.

Residual soil N at the end of the cropping cycle after the treatments with digested and not digested biomasses was greater in B soil than in A soil, particularly after digestate application: we suppose that, in this medium-high fertility soil, N-immobilization process took place at higher extent, probably exerted by the soil microflora which was more abundant in B soil, as confirmed by biochemical parameters reported in Table 1. Also the texture could have played a key-role in this process: the textural characteristics of Soil B (~

84%) probably favoured the ammonium fixation, particularly on the loamy-clay components of that soil.

Soil A	NUE ($N_{uptaken}/N_{fert}$ %)	Residual soil N (g pot⁻¹)	Assimilable N-NO$_3$ (mg pot⁻¹)	Exchangeable N-NH$_4$ (mg pot⁻¹)
Not fertilized	0	0,82 a	5,1 c	21,9 b
Urea 200 Kg ha⁻¹	27,1 d	1,49 b	60,2 d	18,3 a
Urea 400 Kg ha⁻¹	2,3 a	1,75 c	146,1 e	25,8 c
Not digested 200 Kg ha⁻¹	4,6 b	1,61 c	4,8 b	26,9 c
Not digested 400 Kg ha⁻¹	6,1 bc	1,89 c	4,1 a	21,9 b
Digested 200 Kg ha⁻¹	4,1 b	1,99 c	4,3 a	23,6 c
Digested 400 Kg ha⁻¹	8,0 c	1,97 c	4,7 b	22,3 bc
Soil B	NUE ($N_{uptaken}/N_{fert}$ %)	Residual soil N (g pot⁻¹)	Assimilable N-NO$_3$ (mg pot⁻¹)	Exchangeable N-NH$_4$ (mg pot⁻¹)
Not fertilized	0	0,89 a	3,1 a	23,8 ab
Urea 200 Kg ha⁻¹	27,3 d	1,10 b	5,8 c	26,1 b
Urea 400 Kg ha⁻¹	12,8 c	1,23 b	67,1 d	32,5 c
Not digested 200 Kg ha⁻¹	6,8 b	1,43 c	4,2 b	22,5 a
Not digested 400 Kg ha⁻¹	2,4 a	2,44 d	5,8 c	22,1 a
Digested 200 Kg ha⁻¹	2,3 a	2,11 cd	6,5 c	27,4 c
Digested 400 Kg ha⁻¹	2,6 a	2,87 d	6,1 c	25,3 b

Table 3. Nitrogen use efficiency NUE ($N_{uptaken}/N_{fert}$ %), residual soil N (g pot⁻¹), N-NO$_3$ and N-NH$_4$ in relation to the different fertilization treatments in soil A and B (average value; different letters means significant differences at P-level<0.05).

In relation to the N forms available in the soils at the end of the experiment, after the treatments with the organic biomasses, the main mineral N fraction is represented by ammonium form, with the exception of urea application, which dramatically promoted the increase of soil nitrate: this finding is extremely positive, since both the digested and the not-digested solid fraction of livestock manure allowed to limit the excess of nitrate ions in the soil solution. Moreover, it is noticeable that the use of these organic biomasses addressed the N immobilization by the microbial biomass activity, guaranteeing an increase of soil residual fertility.

4. Conclusion

Digested, but also the not digested solid fraction of swine livestock manure, showed N availability compatible with those requested from lettuce, so to be used to fertilize short-term horticultural crops. Obtained lettuce plants had a good quality at the end of the experiment, in both the soils and for both the used organic biomasses. Even if the plant dry matter was partially reduced respect to that obtained after mineral fertilization, the macro and micronutrient uptake was lower after organic biomasses addition, indicating an optimization of nutrient use efficiency. Besides, N uptake was well balanced, so to be the N supply able to guarantee a correct development of lettuce.

Since the main aim of our work was to demonstrate the possibility to use these processed biomasses as N source for plant cultivation by limiting the potential nitrate leaching in soil, as normally happened when livestock manure are directly applied to soil, the obtained results confirmed the significant environmental benefits due to the use of anaerobically digested biomasses as organic fertilizers. They showed to be able to promote the development of lettuce crop, probably through the slow release of available N into the soil system, without incurring in toxicity phenomena also when applied at high N supply, being this positive effect particularly effective in a sandy soil, poor in organic matter.

Thus, the biodigestion process represents a great opportunity not only for generating renewable energy, but also to obtain an organic biomass with good fertilizer/amendment characteristics to be applied to the soil, giving the possibility of use alternative nitrogen sources, contemporary promoting plant growth and limiting the environmental impact.

Author details

Alessandra Trinchera*, Carlos Mario Rivera, Andrea Marcucci and Elvira Rea
*Agricultural Research Council - Research Centre for Plant-Soil System (CRA-RPS).,
Rome, Italy*

Acknowledgement

Many thanks are due to Dr. Sergio Piccinini and Dr. Paolo Mantovi of the "Research Centre for Animal Production - Foundation Centre for Studies and Research" (C.R.P.A. – F.C.S.R. S.p.A., Reggio Emilia, Italy) for their kind support in supplying the organic biomasses applied in this work.

Special thanks are due to Prof. Carlo Grignani of the Department of Agronomy, Forestry and Soil Management (Agroselviter) – University of Turin (Italy) and his staff, for providing soils samples for the following research activities.

5. References

[1] Monreal CM, Dinel H, Schnitzer M, Gamble DS, Biederbeck VO (1997) Impact of carbon sequestration on functional indicatores of soil quality as influenced by management in sustainable agriculture. ISBN: CRC Press LLC. 30: 435-457.

[2] Canali S, Trinchera A, Intrigliolo F, Pompili L, Nisini L, Mocali S, Torrisi B (2004) Effect of long term addition of composts and poultry manure on soil quality of citrus orchards in Southern Italy. Biol. Fert. Soils. 40: 206-210.

[3] Paustian K, Conant R, Ogle S, Paul E (2002) Environmental and management drivers of soil organic C stock changes. Proc. of the OECD Expert Meeting on soil organic C indicators for agricultural land. Ottawa, Canada.

[4] McNeill AM , Eriksen J, Bergstrom L, Smith KA, Marstorp H, Kirchmann H, Nilsson I (2005) Nitrogen and sulphur management: challenges fororganic sources in temperate agricultural systems. Soil Use and Man. 21: 82-93.

* Corresponding Author

[5] Weiland P, Rieger C, Ehrmann T (2003) Evaluation of the newest biogas plants in Germany with respect to renewable energy production, greenhouse gas reduction and nutrient management. Future of Biogas in Europe II, Esbjerg 2-4 October 2003.

[6] Rea E, De Lucia B, Ventrelli A, Pierandrei F, Rinaldi S, Salerno A, Vecchietti L, Ventrelli V (2007) Substrati alternativi a base di compost per l'allevamento in contenitore di specie ornamentali mediterranee. Giornata tematica CIEC "Substrati di coltivazione: sviluppi qualitativi, tecnici, legislativi e commerciali", Milan (Italy), January 18-19 2007.

[7] Sequi P, Rea E, Trinchera A (2007 Aspetti legislativi per la normazione dei substrati di coltivazione. Giornata tematica CIEC "Substrati di coltivazione: sviluppi qualitativi, tecnici, legislativi e commerciali", Milan (Italy), January 18-19 2007.

[8] Lemaire F (1995) Physical, chemical and biological properties of growing medium. Acta Horticol. 396:273-284.

[9] Trinchera A, Benedetti A, Antonelli M, Salvatori S, Nisini L (2007) Organic matter characterisation of amended soils under crop rotation in Mediterranean area. Geoph. Res. Abs. 9: 7635.

[10] Trinchera A, Tittarelli F, Intrigliolo F (2007) Study of organic matter evolution in citrus compost by isoelectrofocusing technique. Comp. Sci. Util. 15(2):101-110.

[11] Trinchera A, Rivera CM, Marcucci A, Rinaldi S, Sequi P, Rea E (2010) Assessing agronomical performances of digested livestock manure to front nitrate leaching in soil. Proceedings of the 18th European Biomass Conference and Exhibition: From Research to Industry and Markets, Lyon, May 3-7 2010. pp.2223-2227

[12] Trinchera A, Allegra.M, Roccuzzo G, Rea E, Rinaldi S, Sequi P, Intrigliolo F (2011) Organo-mineral fertilizers from glass-matrix and organic biomasses. A new way to release nutrients. J. Sci. Food Agr. 91(13): 2386-2393.

[13] Trinchera A, Rivera CM, Rinaldi S, Salerno A, Rea E, Sequi P (2010) Granular size effect of clinoptilolite on maize seedlings growth. Open Agr. J. 4: 23-30.

[14] Council Directive 91/676/EEC concerning the protection of waters against pollution caused by nitrates from agricultural sources for the period 2004-2007.

[15] Thomsen IK,Hansen JF, Kjellerup V , Christensen BT (1997) Effects of cropping system and rates of nitrogen in animal slurry and mineral fertilizer on nitrate leaching from a sandy loam. Soil Use and Man. 9: 53-58.

[16] Tittarelli F, Trinchera A, Intrigliolo F, Benedetti A, (2002) Evaluation of organic matter stability during the composting process of agroindustrial wastes. Microbiology of composting: H Insam, N Riddeck, S Klammer (Eds.) pp. 397-406.

[17] Abdullahi YA, Akunna JC, White NA, Hallett PD, Wheatley R (2008) Investigating the effects of anaerobic and aerobic post-treatment on quality and stability of organic fraction of municipal solid waste as soil amendment. Biores. Technol. 8631–8636.

[18] RodrõÂguez Andara A , Lomas Esteban JM (1999) Kinetic study of the anaerobic digestion of the solid fraction of piggery slurries. Biom. and Bioen. 17: 435-443.

[19] Springer U, Klee J (1954) Prüfung der Leistungsfähigkeit von einigen wichtigeren Verfahren zur Bestimmung des Kohlemstoffs mittels Chromschwefelsäure sowie Vorschlag einer neuen Schnellmethode. Z. Pflanzenernähr. Dang. Bodenk. 64: 1.

[20] Italian Official Methods for Soil Analysis. Official Italian Gazette, 21/10/199, n. 248.

[21] Vance, ED, Brookes PC, Jenkinson DS (1987) An extraction method for measuring microbial biomass C. Soil Biol. Biochem. 19:703-707.

[22] Dommergues Y (1960) La notion de coefficient de minéralisation du carbone dans les sols. Agron. Trop. XV(1):54-60.

[23] Trinchera A, Pinzari F, Benedetti A, Sequi P (1999) Use of biochemical indexes and changes in organic matter dynamics in a Mediterranean environment: a comparison between soils under arable and set-aside management. Org. Geochem. 30:453-459.

[24] Haase DL, Rose R (1995) Vector analysis and its use for interpreting plant nutrient shifts in response to silvicultural treatments. Forest Sci. 41: 56-66.

[25] Swift KI, Brockley RP (1994) Evaluating the nutrient status and fertilization response potential of planted spruce in the interior of British Columbia. Can. J. For. Res. 24: 594-602.

[26] Trinchera A, Pinzari F, Benedetti A (2001) Should we be able to define soil quality before "restoring" it? Use of soil quality indicators in Mediterranean ecosystems. Minerva Biotech. 13: 13-18.

[27] Benedetti A, Baroccio F, Trinchera A (2009) Characterization and efficiency of slow release fertilizers. Proceedings of 16th Nitrogen Workshop, June,28th–July, 1st 2009, Turin (Italy). Grignani C., Acutis M., Zavattaro L., Bechini L., Bertora C., Gallina P.M., Sacco D. (Eds.). pp. 393-394.

[28] Benedetti A, Baroccio F, Trinchera A, Mocali S (2009) Potential mineralization of soil organic nitrogen in aerobic and anaerobic conditions. Proceedings of 16th Nitrogen Workshop, June,28th–July, 1st 2009, Turin (Italy). Grignani C., Acutis M., Zavattaro L., Bechini L., Bertora C., Gallina P.M., Sacco D. (Eds.). pp. 23-24.

[29] Trinchera A, Nardi P, Benedetti A (2010) Influence of biological fertility of soil on methylenurea biodegradation. Proceedings of 18th Symposium of the International Scientific Centre of Fertilizers, 8-12 November 2009, Rome (Italy). pp. 506-510.

[30] Govi M, Ciavatta C, Gessa C (1994) Evaluation of the stability of the organic matter in slurries, sludges and composts using humification parameters and isoelectric focusing. Humic Substances in the Global Environment and Implications on Human Health. Senesi S and Miano TM (Eds). Elsevier Science. pp. 1311-1316.

[31] Govi M, Ciavatta C, Montecchio D, Sequi P (1995) Evolution of organic matter during stabilization of sewage sludge. Agr. Med., 125: 107-114.

[32] Valentine DW, Allen HL (1990) Foliar responses to fertilization identify nutrient limitation in loblolly pine. Can. J. For. Res. 20: 144-151.

[33] Scagel CF (2003) Growth and nutrient use of ericaceous plants grown in media amended with sphagnum moss peat or coir dust. Soil Management, fertilization and irrigation. HortSci. 38(1): 46-54.

Chapter 2

Animal Manures:
Recycling and Management Technologies

María Gómez-Brandón, Marina Fernández-Delgado Juárez,
Jorge Domínguez and Heribert Insam

Additional information is available at the end of the chapter

1. Introduction

Many environmental problems of current concern are due to the high production and local accumulations of organic wastes that are too great for the basic degradation processes inherent in nature. With adequate application rates, animal manure constitutes a valuable resource as a soil fertilizer, as it provides a high content of macro- and micronutrients for crop growth and represents a low-cost, environmentally- friendly alternative to mineral fertilizers [1]. However, the intensification of animal husbandry has resulted in an increase in the production of manure - over 1500 million tonnes are produced yearly in the EU-27 [2] as reported by Holm-Nielsen et al. [3]- that need to be efficiently recycled due to the environmental problems associated with their indiscriminate and untimely application to agricultural fields. The potentially adverse effects of such indiscriminate applications include an excessive input of harmful trace metals, inorganic salts and pathogens; increased nutrient loss, mainly nitrogen and phosphorus, from soils through leaching, erosion and runoff-caused by a lack of consideration of the nutrient requirements of crops; and the gaseous emissions of odours, hydrogen sulphide, ammonia and other toxic gases [4]. In fact, the agricultural contribution to total greenhouse gas emissions is around 10%, with livestock playing a key role through methane emission from enteric fermentation and through manure production. More specifically, around 65% of anthropogenic N_2O and 64% of anthropogenic NH_3 emissions come from the worldwide animal production sector [5].

The introduction of appropriate management technologies could thus mitigate the health and environmental risks associated with the overproduction of organic wastes derived from the livestock industry by stabilizing them before their use or disposal. Stabilisation involves the decomposition of an organic material to the extent of eliminating the hazards and is normally reflected by decreases in microbial biomass and its activity and in concentrations of labile compounds [6]. Composting and vermicomposting have become two of the best-known environmentally appropriate technologies for the recycling of manures under aerobic conditions [6-7], by transforming them into safer and more stabilised products (compost and vermicompost) with benefits for both agriculture and the environment. Unlike composting, vermicomposting depends on the joint action between earthworms and microorganisms and does not involve a thermophilic phase [8]. However, more than a century had to pass before vermicomposting was truly considered a field of scientific knowledge or even a real technology, despite Darwin [9] having already highlighted the important role of earthworms in the decomposition of dead plants and the release of nutrients from them.

Although microbial degradation under oxygen is usually faster and, as such aerobic processes are thermodynamically more favorable than anaerobic processes, in recent years, anaerobic digestion (AD) has become an upcoming technology for the treatment of animal manures [3, 10-13]. On the one hand, pretreatment of manure by anaerobic digestion can involve some advantages including malodor reduction, decreased biochemical oxygen demand, pathogen control, along with a reduction in the net global warming potential of the manure [4,14]. AD reduces the risk of water pollution associated with animal manure slurries (i.e., eutrophication) by removing 0.80–0.90 of soluble chemical oxygen demand and it improves human/farm cohabitation in rural regions by reducing odor emissions by 70–95% [4]. This process has other direct advantages beyond these, which are related to biogas production for renewable energy and the enrichment of mineral fractions of N and P during digestion [4,10], resulting in a more balanced nutrient mix and increased nutrient bioavalability for plants compared with undigested manure [15].

Therefore, the purpose of this chapter is to give an overview of the three major management technologies of manure recycling, including the aerobic processes of composting and vermicomposting and the anaerobic digestion for biogas production. The main changes that occur in the substrate from a chemical and microbial viewpoint during the specific phases of each degradation process are addressed, as such changes determine the degree of stability of the end product and in turn its safe use as an organic amendment. Different methods that have been proposed to evaluate compost stability are summarised. Also, the influence of the end products derived from each process on the soil microbiota and disease suppressiveness are discussed.

2. Aerobic degradation: Composting and vermicomposting processes

Under aerobic conditions, the degradation of organic matter is an exothermic process during which oxygen acts as a terminal electron acceptor and the organic materials are transformed into more stable products, carbon dioxide and water are released, and heat is evolved. Under field conditions, aerobic degradation takes place slowly at the soil surface, without reaching high temperatures; but this natural breakdown process can be accelerated by heaping the material into windrows to avoid heat losses and thus allowing for temperature increases (composting) or by using specific species of earthworms as agents for turning, fragmentation, and aeration (vermicomposting). Although both aerobic processes, composting and vermicomposting, have been widely used for processing different types of animal manure either separately or in combination with each other (see Table 1), most of the studies are not comparable mainly due to differences in the applied experimental designs, parent material, earthworm species, as well as the length of the experiments and the parameters used for analysis, among others. Despite these limitations, all these findings have largely contributed to better understand the changes that the material undergoes during these biological stabilisation processes, which is of great importance for their optimisation, and ultimately to obtain a high quality final product. In line with this, certain chemical characteristics of the animal manures can limit the efficiency of these processes, such as an excess of moisture, low porosity, a high N concentration in relation to the organic C content or high pH values [6]. Therefore, different aeration strategies, substrate conditioning-feedstock formulation, bulking agents and process control options have been considered in manure composting and vermicomposting so as to reduce the time and costs of both processes and enhance the quality of the end-products [6-7].

2.1. The composting process

Composting is defined as a bio-oxidative process involving the mineralization and partial humification of the organic matter, leading to a stabilised final product, free of phytotoxicity and pathogens and with certain humic properties, which can be used to improve and maintain soil quality and fertility [25]. Composting of animal manures has been traditionally carried out by the farmers after manure collection for better handling, transport and management [6]. Frequently, the wastes were heaped up and very little attention was paid to the process conditions (aeration, temperature, ammonia loss, etc.) and using rudimentary methodology.

From a microbial viewpoint continuous composting processes may be described as a sequence of continuous cultures, each of them with their own physical (temperature), chemical (the available substrate), and biological (i.e., the microbial community composition) properties and feedback effects. These changes make it difficult to study the process, which is virtually impossible to simulate in the laboratory since temperature, moisture, aeration, etc., are directly related to the surface/volume ratio. However, in general, composting may be described as a four-phase process in which the energy-rich, abundant and easily degradable compounds like sugars and proteins are degraded by fungi and bacteria (referred to as primary decomposers) during an initial phase called the *mesophilic phase* (25-40 °C). Although there exists a competition between both microbial groups regarding the easily available substrates, fungi are very soon outcompeted because the maximum of specific growth rates of bacteria exceed those of fungi by one order of magnitude [26]. The importance of bacteria (with the exception of Actinobacteria) during the composting process has long been neglected, probably because of the better visibility of mycelial organisms. A review on the microbial groups involved in the first mesophilic phase is given by [27]. Provided that mechanical influences (like turning) are small, compost fauna including earthworms, mites and millipedes may also act as catalysts, thereby contributing to the mechanical breakdown and offering an intestinal habitat for specialized microorganisms. The contribution of these animals may be negligible or, as in the special case of vermicomposting, considerable (see section 2.2). The number of mesophilic organisms in the original substrate is three orders of magnitude higher than the number of thermophilic organisms; however, the activity of primary decomposers induces a temperature rise and in turn, mesophilic microbiota is, along with the remaining easily degradable compounds, degraded by the succeeding thermophiles. The temperature rise continues to be fast and accelerates up to a temperature of about 62 °C during this second phase of composting, known as the *thermophilic phase*.

When a temperature exceeding 55 °C is reached in a compost pile, fungal growth is usually inhibited and the thermophilic bacteria and Actinobacteria are the main degraders during this peak-heating phase. Moreover, oxygen supply affects fungi to a greater extent than bacteria, and even in force-aerated systems, temporary anoxic conditions may occur. Hence, fungi play a negligible role during this phase, except for the composting of lignocellulosic residues. Bacteria of the genus *Bacillus* are often dominant when the temperature ranges from 50 to 65 °C. Moreover, members of the *Thermus/Deinococcus* group have been found in biowaste composts [28] with an optimum growth between 65 and 75 °C. A number of autotrophic bacteria that obtain their energy by the oxidation of sulfur or hydrogen have been isolated from composts [28]. Their temperature optimum is at 70-75 °C and they closely resemble *Hydrogenobacter* strains, which were previously found in geothermal sites. Furthermore, obligate anaerobic bacteria are also common in composts, but up to now, there is still a gap of knowledge concerning this microbial group. It is believed that the longer generation times of archaea, in comparison with bacteria, made the archaea unsuitable for the rapidly changing conditions in the composting process. Nevertheless, in recent works, and using the right tools, a considerable number of cultivable (*Methanosarcina termophila*, *Methanothermobacter* sp., *Methanobacterium formicicum*, among others) and yet uncultivated archaea have been detected in composting processes [29-30].

The final temperature increase may exceed 80 °C and it is mainly due to the effect of abiotic exothermic reactions in which temperature-stable enzymes of Actinobacteria might be involved. Such high temperatures are crucial for compost hygienisation in order to destroy human and plant pathogens, and kill weed seeds and insect larvae [31]. The disadvantage of temperatures exceeding 70 °C is that most mesophiles are killed, and therefore the recovery of the decomposer community is retarded after the temperature peak. The inoculation with matter from the first mesophilic stage might, however, solve this problem.

When the activity of thermophilic organisms ceases due to the exhaustion of substrates, the temperature starts to decrease. This constitutes the beginning of the third stage of composting, called the *cooling phase* or *second mesophilic phase*. It is characterised by the recolonisation of the substrate with mesophilic organisms, either originating from surviving spores, through the spread from protected microniches, or from external inoculation. During this phase there is an increased number of organisms with the ability to degrade cellulose or starch, such as the bacteria *Cellulomonas*, *Clostridium* and *Nocardia*, and fungi of the genera *Aspergillus*, *Fusarium* and *Paecilomyces* [27]. Finally, during the *maturation phase*, the ratio of fungi to bacteria increases due to the competitive advantage of fungi under conditions of decreasing water potential and poorer substrate availability. Compounds that are not further degradable, such as lignin-humus complexes, are formed and become predominant. Some authors have proposed a fifth composting phase, known as the *curing phase* (or *storage phase*), during which the physico-chemical parameters do not change, but changes in microbial communities still occur [32]. Therefore, the chemical and microbial changes that the substrate undergoes during the different phases of the composting process will largely determine the stability and degree of maturity of the end product and in turn, its safe use as an organic amendment. There exists a wide range of parameters that have been proposed to evaluate compost stability/maturity, as shown in the next section.

2.1.1. Evaluation of compost stability and maturity

The stability and maturity of compost is essential for its successful application, particularly for composts used in high value horticultural crops [33]. Both terms are usually used interchangeably to describe the degree of decomposition and transformation of the organic matter in compost [34], despite the fact that they describe different properties of the composting substrate. Stability is strongly related to the degree to which composts have been decomposed to more stable organic materials [35]. Unstable compost, in contrast, contains a high proportion of biodegradable matter that may sustain a high microbial activity [36]. Typically, compost stability is evaluated by different respirometric measurements and/or by studying the transformations in the chemical characteristics of compost organic matter [6]. On the other hand, compost maturity generally refers to the degree of decomposition of phytotoxic organic substances produced during the active composting stage and to the absence of pathogens and viable weed seeds [37]. This property is often characterised by germination indexes [38] and/or by nitrification [6] and has been related to the degree of compost humification. Wu et al. [37] reported that the low CO_2 evolution is not always an indicator of a non-phytotoxic compost, which suggests that a stable compost may not always be at a level of maturity suitable for its use as a growing medium for certain species of plant.

Several authors highlight that there is no one single method that can be applied successfully for determining compost stability mainly due to the wide range of raw materials used to produce compost, as reviewed by [6]. Therefore, the integration of different parameters seems to be the most reliable option for evaluating the stability/maturity stage of composted materials. For instance, physical parameters including temperature, odor and color constitute a very simple and rapid method for stability evaluation, giving a general idea of the decomposition stage reached; however, little information is achieved as regards to the degree of maturation. In addition, chemical parameters including pH, electrical conductivity, cation exchange capacity (CEC), the ratios of C to N and NH_4^+ to NO_3^- and humification parameters have also been widely used as indicators of stability [39]. Nevertheless, several drawbacks have been found regarding these parameters, thereby preventing their use as accurate indicators. According to Wu et al. [37], pH and electrical conductivity may be used to monitor compost stabilisation, as long as the source waste composition is relatively consistent and other stability tests are conducted. Moreover, Namkoong et al. [40] established that the C to N ratio could not be considered as a reliable index of compost stability, as it changed irregularly with time. In fact, when wastes rich in nitrogen are used as the source material for composting, like sewage sludges or manures, the C to N ratio can be within the values of a stable compost even though it may still be unstable. Zmora-Nahum et al. [34] reported a C to N ratio lower than the cutoff value of 15 very early during the composting of cattle manure, while important stabilisation processes were still taking place. The increase in CEC with composting time is related to the formation of carboxyl and phenolic functional groups during the humification processes; however, the wide variation in CEC values among the initial substrates prevent to establish a threshold level and to use it as a stability indicator [41].

Biological parameters such as respiration rates (CO_2 evolution rate and/or O_2 uptake rate) and enzyme activities have been proposed to measure compost stability [6-39]. The principle of the respirometric tests is that unstable compost has a strong demand for O_2 and high CO_2 production rates as a consequence of the intensive microbial development due to the presence of easily biodegradable compounds in the raw material. Then, as composting proceeds, the decrease in the amount of degradable organic matter is accompanied by a decline in both O_2 and CO_2 respirometry. The Solvita test, which measures CO_2 evolution and ammonia emissions simultaneously have been found to be a simple and easily used procedure for quantifying soil microbial activity in comparison with both titration and infrared gas analysis [42]; this test has also been used for determining the stability degree in diverse composts [43]. Enzymatic activities have also been found suitable as indicators of the state and evolution of the organic matter during composting, as they are implicated in the biological and biochemical processes through which the initial organic substrates are transformed (Tiquia, 2005). Important enzymes during composting are related to the C-cycle (cellulases, β-glucosidase, β-galactosidase), the N-cycle (protease, urease, amidase) and/or the P-cycle (phosphatase) [44]. These latter authors established that the formation of a stable enzymatic complex, either in moist or air-dried compost samples, could represent a useful index of stabilisation. Additionally, enzymatic activities, especially dehydrogenases, are considered easy, quick and cheap stability measurements [36]; however, the wide range of organic substrates involved in the composting process makes it difficult to establish general threshold values for these parameters. The hydrolysis of fluorescein diacetate (FDA), which is a colourless fluorescein conjugated that is hydrolyzed by both free (exoenzymes) and membrane bound enzymes [45], has been suggested as a valid parameter for measuring the degree of biological stability of the composting material, as it showed a good correlation with other important stability indexes [46]. The analysis of phospholipid fatty acid (PLFA) composition has also been proposed for determining compost stability [47-48]. These authors found a positive correlation between the proportion of PLFA biomarkers for Gram-positive bacteria and the germination index during the maturation of composting of poultry manure and cattle manure, respectively. The strength of this lipid-based approach, as compared to other microbial community assays, is that PLFAs are rapidly synthesized during microbial growth and quickly degraded upon microbial death and they are not found in storage molecules, thereby providing an accurate 'fingerprint' of the current living community [49]. The potential ability of the microbial community to utilise select carbon sources by determining the community-level physiological profiles (CLPPs) with the Biolog® Ecoplate has also been considered for compost stability testing [50]. The principle is that compost extracts are inoculated onto microtiter plates that contain 31 different C substrates [51-52]. Ultimately, molecular techniques are becoming increasingly useful in composting research. For example, as in reference [32] the authors used three different cultivation independent techniques based on 16S rRNA gene sequences, i.e. PCR-denaturing gradient gel electrophoresis (DGGE), clone libraries, and an oligonucleotide microarray (COMPOCHIP), in order to evaluate the dynamics of microbial communities during the compost-curing phase. Specific compost-targeted microarrays are suitable to investigate bacterial [53-54] and fungal community patterns [55], including plant growth promoting organisms and plant and human pathogens.

2.1.2. Influence of compost amendments on the soil microbiota

Composted materials have gained a wide acceptance as organic amendments in sustainable agriculture, as they have been shown to provide numerous benefits whereby they increase soil organic matter levels, improve soil physical properties (increased porosity and aggregate stability and reduced bulk density) and modify soil microbial communities [56]. Substantial evidence indicates that the use of compost amendments typically promotes an increase in soil microbial biomass and activity, as reviewed in [56-57]. This enhancing effect may be attributed to the input of microbial biomass as part of the amendments [58]; however, the quantity of organic matter applied with the compost is very small in comparison with the total organic matter present in the soil and, in turn it is believed that the major cause is the activation of the indigenous soil microbiota by the supply of C-rich organic compounds contained in the composting materials [58]. Such effects on microbial communities were reported to be dependent on the feedstocks used in the process [59]. However, other authors did not find significant differences between soil plots that had been amended with four different compost types (green manure compost, organic waste compost, manure compost and sewage sludge compost) over 15 years [60]. Similarly, Ros et al. [61] observed that different types of composts had a similar effect on the fungal community and microbial biomass in soil in a long-term field experiment. This fact suggests that the soil itself influences the community diversity more strongly than the compost treatments. Such discrepancies between the previous findings may be due to differences in soil properties, land-use and compost type (i.e., different starting material and process parameters), frequency and dose of application, length of the experiment and parameters chosen for analysis, among others.

Furthermore, C addition to soil seems to select for specific microbial groups that feed primarily on organic compounds. Therefore, it can be expected that the addition of organic amendments not only increases the size of the microbial community but also changes its composition, as has already been observed in previous experiments [61-63]. As shown by [62] higher amounts of composts resulted in a more pronounced and faster effect in the structure of microbial communities, as revealed by PLFA analysis, indicating that the compost application rate is a major factor regarding the impact of compost amendments on soil microbiota. Carrera et al. [63] found that soil PLFA profiles were influenced by both the treatment with poultry manure compost and the sampling date. Ros et al. [61] observed that the date of sampling contributed more to modifications in fungal community structure, assessed by PCR-DGGE analysis, than treatment effects. However, in contrast to fungi, the bacterial community structure, both on the universal and the Streptomycetes group-specific level, were influenced by compost amendments, especially the combined compost and mineral fertilisers treatments. This seems reasonable, as bacteria have a much shorter turnover time than fungi and can react faster to the environmental changes in soil. Bacterial growth is often limited by the lack of readily available C substrates, even in soils with a high C/N ratio, and are the first group of microorganisms to assimilate most of the readily available organic substrates after they are added to the soil [64]. CLPP profiles have also been used to evaluate the impact of compost amendments on the potential functional diversity of soil microbiota, as they are considered suitable indicators for detecting soil management changes [65]. As shown by [66] different types of compost (household solid waste compost and manure compost) affected differently the substrate utilization patterns of the soil microbial community relative to unamended control soils. Contrarily, no significant changes in CLPP profiles were found by [59]. Other authors also reported that the sampling date had more weight on CLPP results than compost treatments [63,67]. All these studies together highlight the importance of a multi-parameter approach for determining the influence of compost amendments on the soil microbiota, which is of utmost importance to understand the disease suppressive activity of compost and the mechanisms involved in such suppression [68].

Since the 1980s a large number of experiments have been addressed describing a wide array of pathosystems and composts from a broad variety of raw materials. Interestingly, Noble and Coventry [69] evaluated the suppression of soilborne plant pathogens by compost in both laboratory and field scale experiments. In general, they found that the effects in the field were smaller and more variable than those observed at lab-scale. Termorshuizen et al. [70] compared the effectiveness of 18 different composts on seven pathosystems and interestingly they found significant disease suppression in 54% of the cases, whereas only 3% of the cases showed significant disease enhancement. They highlighted that the different composts did not affect the pathogens in the same way and that no single compost was found to be effective against all the pathogens. Furthermore, in a study carried out with 100 composts produced from various substrates under various process conditions, it was found that those composts that had undergone some anaerobic phase showed the best results in terms of suppressing plant disease [71].

However, up to now, there is still a general lack of understanding concerning the suppressivity of compost [68], as it depends on a complex range of abiotic and biotic factors. Such factors are reviewed by [72]. Briefly, the main mechanisms by which compost amendments exert their suppressivity effect against soil-borne plant pathogens include hyperparasitism; antibiosis; competition for nutrients (carbon and/or iron); and induced systemic resistance in the host plant [73]. The first three affect the pathogen directly and reduce its survival, whilst the latter one acts indirectly via the plant and affect the disease cycle.

Manure type	Bulking agent	Composting process	Vermicomposting process	Duration of experiment	Investigated parameters	Remarks	References
Pig manure	-	-	Vertical continuous-feeding system (E. fetida)	36 weeks	Microbial biomass C, basal respiration, metabolic quotient (qCO₂), CLPPs, enzymatic activities	Increase in functional microbial diversity with earthworm presence associated with a decrease in qCO₂, indicative of high metabolic efficiency

Increase in cellulase activity with earthworm presence accompanied by greater C losses throughout the process | [16-17] |
| Pig manure | - | - | Vertical continuous-feeding system (E. fetida) | 36 weeks | PLFAs, basal respiration | Decrease in bacterial and fungal biomass, assessed by PLFA biomarkers, associated with an increase in microbial activity in the presence of earthworms | [18] |
| Sheep manure | Olive waste | Turned windrow | Vermicomposting bed (E. fetida) | 36 weeks for both composting and vermicomposting | Enzymatic activities, PCR-DGGE, real time-PCR | Higher bacterial abundance and microbial diversity was found in vermicompost relative to the initial substrate than in compost | [19] |

Manure type	Bulking agent	Composting process	Vermicomposting process	Duration of experiment	Investigated parameters	Remarks	References
Cow manure	Straw	Forced-ventilation	Vermicomposting bed (E. andrei)	Composting: 15 d (thermophilic phase) Vermicomposting: 40 d (active phase)	Microbial biomass C, basal respiration, enzymatic activities, ergosterol	Lower levels of microbial biomass and dehydrogenase activity indicative of a higher degree of stabilisation were found in the combined treatment (composting + vermicomposting)	[20]
Cow manure	Agricultural plant waste	In-vessel system	Semicontinuously vermicomposting system (E. fetida)	Composting: 3-4 weeks (thermophilic phase) Vermicomposting: not specified	Substrate-induced respiration, PCR-DGGE, CompoChip	Vermicompost tea composition was influenced by production and storage conditions. Carbon substrate addition during the tea production process was identified to be of major importance for obtaining a vermicompost tea with a rich and diverse microbial community	[21]
Cow manure+poultry manure	Biochar	Turned windrow	-	12 weeks	PLFAs	Changes in microbial community composition depended on the origin of manure composted with biochar	[22]
Cow manure	Sawdust	Turned windrow	-	30 d	PCR-DGGE, clone libraries	Ammonia oxidizing archaea were found to be dominant during the composting process probably because they could adapt to increasing temperature and/or nutrient loss	[23]
Cow manure+Horse manure+Pig Manure	-	-	In-container system (E. andrei, E. fetida and P. excavatus)	30 d (active phase)	Basal respiration, microbial growth rates, PLFAs	Species-specific effects of earthworms on microbial community structure and bacterial growth rate	[24]

Table 1. Research studies focused on the stabilization of animal manures through the aerobic processes of composting and vermicomposting.

Two mechanisms of biological control, based on antibiosis, hyperparasitism, competition and induced protection, have been reported for compost amendments. On the one hand, diseases caused by plant pathogens such as *Phytophtora* spp. and *Phytium* spp. have been eradicated through a mechanism known as "general suppression", in which the suppressive activity is attributed to a diverse microbial community in the compost rather than to a population of a single defined species augmented to infested soil. Whilst, for *Rhizoctonia rolani*, few microorganisms present in compost are able to eradicate this pathogen and, in turn this type of suppression is referred as "specific suppression". Overall, all of the above reinforces that the activity of microbial communities in composts is a major factor affecting the suppression of soilborne plant pathogens. Indeed, the disease suppressive effect is usually lost following compost sterilization or pasteurization [68]. Better understanding of the microbial behaviour and structure of the antagonistic populations in the compost will provide tools to reduce its variability. In line with this, Danon et al. [32] detected, using PCR-based molecular methods, distinctive community shifts at different stages of prolonged compost curing being Proteobacteria the most abundant phylum in all the stages, whereas Bacteroidetes and Gammproteobacteria were ubiquitous. Actinobacteria were dominant during the mid-curing stage, and no bacterial pathogens were detected even after a year of curing.

The addition of antagonistic microorganisms to compost is also a promising technique to improve its suppressivity. Already in 1983, Nelson et al. [74] increased the suppressive potential of compost by adding selected *Trichoderma* strains. They found that not only the addition of the antagonist is important, but also the strategy of inoculation of the antagonist in order to efficiently colonize the substrate, as the autochthonous microbial community can inhibit it. Ultimately, predicting disease suppression on the basis of pure compost is expected to be highly advantageous for compost producers. This would enable them to optimise the composting process based on the specific disease jeopardising the target crop. For this purpose, a further step could be the development of quality control criteria based mainly on bioassays designed for a specific pathogen or disease.

2.2. The vermicomposting process

Vermicomposting is defined as a bio-oxidative process in which detritivore earthworms interact intensively with microorganisms and other fauna within the decomposer community, accelerating the stabilization of organic matter and greatly modifying its physical and biochemical properties [75]. Epigeic earthworms with their natural ability to colonize organic wastes; high rates of consumption, digestion and assimilation of organic matter; tolerance to a wide range of environmental factors; short life cycles, high reproductive rates, and endurance and resistance to handling show good potential for vermicomposting [76]. The earthworm species *Eisenia andrei*, *Eisenia fetida*, *Perionyx excavatus* and *Eudrilus eugeniae* display all these characteristics and they have been extensively used in vermicomposting facilities.

Vermicomposting systems sustain a complex food web that results in the recycling of organic matter and release of nutrients [77]. Biotic interactions between decomposers (i.e., bacteria and fungi) and earthworms include competition, mutualism, predation and facilitation, and the rapid changes that occur in both functional diversity and in substrate quality are the main properties of these systems [77]. The biochemical decomposition of organic matter is primarily accomplished by the microbes, but earthworms are crucial drivers of the process as they may affect microbial decomposer activity by grazing directly on microorganisms [78-79], and by increasing the surface area available for microbial attack after the comminution of organic matter [8]. Furthermore, earthworms are known to excrete large amounts of casts, which are difficult to separate from the ingested substrate [8]. The contact between worm-worked and unworked material may thus affect the decomposition rates [80], due to the presence of microbial communities in earthworm casts different from those contained in the material prior to ingestion [81]. In addition, the nutrient content of the egested materials differ from that in the ingested material [82], which may enable better exploitation of resources, because of the presence of a pool of readily assimilable compounds in the earthworm casts. Therefore, the decaying organic matter in vermicomposting systems is a spatially and temporally heterogeneous matrix of organic resources with contrasting qualities that result from the different rates of degradation that occur during decomposition [83].

The impact of earthworms on the decomposition of organic waste during the vermicomposting process is initially due to *gut- associated processes* (GAPs), i.e., via the effects of ingestion, digestion and assimilation of the organic matter and microorganisms in the gut, and then casting [79] (Figure 1). Specific microbial groups respond differently to the gut environment [84] and selective effects on the presence and abundance of microorganisms during the passage of organic material through the gut of these earthworm species have been observed. For instance, some bacteria are activated during the passage through the gut, whereas others remain unaffected and others are digested in the intestinal tract and thus decrease in number [78]. These findings are in accordance with a recent work that provides strong evidence for a bottleneck effect caused by worm digestion (*E. andrei*) on microbial populations of the original material consumed [79]. This points to the earthworm gut as a major shaper of microbial communities, acting as a selective filter for microorganisms contained in the substrate, thereby favouring the existence of a microbial community specialised in metabolising compounds produced or released by the earthworms, in the egested materials. Such selective effects on microbial communities as a result of gut transit may alter the decomposition pathways during vermicomposting, probably by modifying the composition of the microbial communities involved in decomposition, as microbes from the gut are then released in faecal material where they continue to decompose egested organic matter. Indeed, as mentioned before, earthworm casts contain different microbial populations to those in the parent material, and as such it is expected that the inoculum of those communities in fresh organic matter promotes modifications similar to those found when earthworms are present, altering microbial community levels of activity and modifying the functional diversity of microbial populations in vermicomposting systems [80]. Previous studies have already shown that a higher microbial diversity exists in vermicompost relative to the initial substrate [19,85]. Upon completion of GAPs, the resultant earthworm casts undergo *cast- associated processes* (CAPs), which are more closely related to ageing processes, the presence of unworked material and to physical modification of the egested material (weeks to months; Figure 1). During these processes the effects of earthworms are mainly indirect and derived from the GAPs [17].

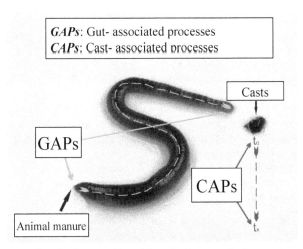

Figure 1. Earthworms affect the decomposition of the animal manure during vermicomposting through ingestion, digestion and assimilation in the gut and then casting (*gut- associated processes*); and *cast-associated processes,* which are more closely related with ageing processes.

Overall, vermicomposting includes two different phases regarding earthworm activity: (i) an *active phase* during which earthworms process the organic substrate, thereby modifying its physical state and microbial composition [86], and (ii) a *maturation phase* marked by the displacement of the earthworms towards fresher layers of undigested substrate, during which the microorganisms take over the decomposition of the earthworm-processed substrate [17-18]. The length of the maturation phase is not fixed, and depends on the efficiency with which the active phase of the process takes place, which in turn is determined by the species and density of earthworms, and the rate at which the residue is applied [8]. During this aging, vermicompost is expected to reach an optimum in terms of its nutrient content and pathogenic load, thereby promoting plant growth and suppressing plant diseases [8]. However, unlike composting, vermicomposting is a mesophilic process (<35 °C), and as such substrates do not undergo thermal stabilisation that eliminates pathogens. Nevertheless, it has been shown that vermicomposting may reduce the levels of different pathogens such as *Escherichia coli, Salmonella enteritidis,* total and faecal coliforms, helminth ova and human viruses in different types of waste [75]. In a recent work [78], a reduction by 98% in the number of faecal coliforms of pig slurry was detected after two weeks of processing in the presence of *E. fetida,* which indicates that the own earthworm digestive abilities play a key role in the reduction of the pathogenic load of the parent material. In a previous study, these authors found that the decrease in pathogenic bacteria (i.e. total coliforms) as a result of gut transit differed among four vermicomposting earthworm species (*Eisenia fetida, Eisenia andrei, Lumbricus rubellus* and *Eudrilus eugeniae*)

[87]. This was consistent with the fact that specific microbial groups respond differently to the gut environment, depending on the earthworm species. The pathogen considered is another important factor controlling the reduction in the pathogenic load during the process. Parthasarathi et al. [88] observed that earthworms did not reduce the numbers of *Klebsiella pneumoniae* and *Morganella morganii*, whereas other pathogens such as *Enterobacter aerogenes* and *Enterobacter cloacae* were completely eliminated. In a recent study [89] a decrease in the abundance of faecal enterococci, faecal coliforms and *Escherichia coli* was recorded across the layers of an industrial-scale vermireactor fed with cow manure; whereas no changes were reported for total coliforms, *Enterobacteria* or *Clostridium*. These findings are of great importance for the optimisation of the vermicomposting process because despite the pioneering studies of Riggle [90] and Eastman et al. [91], little is known about this process in industrial-scale systems, that is, vermicomposting systems designed to deal with large amounts of wastes. This selective effect on pathogens indicates that earthworms not only modify the abundance of such pathogenic bacteria but also alter their specific composition. According to [89], the unaffected pathogens could benefit as a result of the overall decrease in bacterial and fungal biomass across the layers of the reactor, thereby diminishing possible competition for resources.

Collectively, the aforementioned studies highlight the importance of monitoring the changes in microbial communities during vermicomposting, because if the earthworms were to stimulate or depress microbiota or modify the structure and activity of microbial communities, they would have different effects on the decomposition rate of organic matter, thereby influencing the vermicompost properties, which is critical to guarantee a safe use of this end-product as an organic amendment and thus benefit both agriculture and the environment.

2.2.1. Effects of earthworms on microbial communities during vermicomposting: a case study.

Animal manures are microbe-rich environments in which bacteria constitute the largest fraction (around 70% of the total microbial biomass as assessed by PLFA analysis), with fungi mainly present as spores [24]. Thus, earthworm activity is expected to have a greater effect on bacteria than on fungi in these organic substrates in the short-term [79]. In line with this, a significant increase in the fungal biomass of pig manure, measured as ergosterol content, was detected in a short-term experiment (72 h) with the earthworm species *E. fetida*, and the effect depended on the density of earthworms [82]. A higher fungal biomass was found at intermediate and high densities of earthworms (50 and 100 earthworms per mesocosm, respectively), which suggests that there may be a threshold density of earthworms at which fungal growth is triggered. This priming effect on fungal populations was also observed in previous short-term experiments in the presence of the epigeic earthworms *Eudrilus eugeniae* and *Lumbricus rubellus* fed with pig and horse manure, respectively [16,86]. These contrasting short-term effects on bacterial and fungal populations are thus expected to have important implications on decomposition pathways during vermicomposting because important differences exist between both microbial decomposers related to resources requirements and exploitation [92]. This is based on the fact that fungi can immobilise great quantities of nutrients in their hyphal networks, whereas bacteria are more competitive in the use of readily decomposable compounds and have a more exploitative nutrient use strategy by rapidly using newly produced labile substrates [92].

The above-mentioned studies dealing with the effects of epigeic earthworms on microorganisms have focused on the changes before and after the active phase rather than those that occur throughout the whole vermicomposting process. Hence, in a current research study, and using a continuous-feeding vermicomposting system, we evaluated the different phases of interaction between earthworms and microorganisms and additionally, we monitored the stabilisation of the fresh manure during a period of 250 days. At the end of the experiment we obtained a profile of layers of increasing age, resembling a time profile, with a gradient of fresh-to-processed manure from the top to the bottom. This type of system allowed us to evaluate whether and when the samples reached an optimum value to be classified as vermicompost, as regards to the stabilisation of organic matter and the levels of microbial biomass and activity. Briefly, we used polyethylene reactors (n=5) with a volume of 1 m^3, which were initially comprised of a 10 cm layer of mature vermicompost (a stabilised non- toxic substrate that serves as a bed for earthworms), on which earthworms (*Eisenia fetida*) were placed and a layer containing 5 kg of fresh rabbit manure, which was placed over a plastic mesh (5 mm pore size) to avoid sampling the earthworm bedding. New layers with the same amount of fresh manure were added to the vermireactor every fifty days according to the feeding activity of the earthworm population. This procedure allowed the addition of each layer to be dated within the reactors. The reactors were divided into 4 quadrants and two samples were taken at random from each quadrant with a cylindrical corer (8 cm diameter). Each corer sample was divided into five layers of increasing age and the samples from the same layer and each reactor were gently mixed to analyse the changes in microbial communities. The structure of the microbial communities was assessed by PLFA analysis; some specific PLFAs were used as biomarkers to determine the presence and abundance of specific microbial groups [93]. The sum of PLFAs characteristic of Gram-positive (iso/anteiso branched-chain PLFAs), and Gram-negative bacteria (monounsaturated and cyclopropyl PLFAs) were chosen to represent bacterial PLFAs, and the PLFA 18:2ω6c was used as a fungal biomarker. Total microbial activity was also assessed by measuring the rate of evolution of CO_2, as modified for [17] for samples with a high organic matter content. Dissolved organic carbon was determined colorimetrically in microplates after moist digestion ($K_2Cr_2O_7$ and H_2SO_4) of aliquots of 0.5M K_2SO_4 extracts.

The earthworm species *E. fetida* had a strong effect in the decomposition of organic matter during vermicomposting, greatly modifying the structure of the microbial decomposer communities, as revealed by the phospholipid fatty acid analysis. The principal component analysis of the 27 identified PLFAs (10:0, 12:0, 13:0, 14:0, i14:0, 15:0, i15:0, a15:0, 16:0, i16:0, 17:0, a17:0, 18:0, 14:1ω5c, 15:1ω5c, 16:1ω7c, 17:1ω7c, 18:1ω7c, 18:1ω9c, 18:1ω9t, 18:2ω6c, 18:2ω6t, 18:3ω6c, 18:3ω3c, cy17:0, cy19:0, 20:0) clearly differentiated between the samples in function of the age of layers, explaining 51% of the variance in the data (Figure 2). Thus, the upper layers (50 and 100 days old) along with the fresh manure were clearly distinguished from the intermediate (150 days old) and lower layers (200 and 250 days old) (Figure 2).

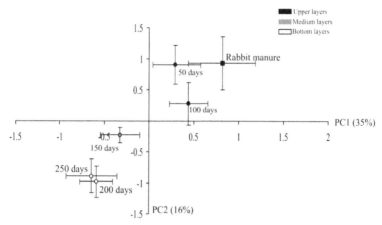

Figure 2. Changes in the microbial community structure throughout the process of vermicomposting assessed by the principal component analysis of the twenty-seven PLFAs identified in the layers of reactors fed with rabbit manure. The different layers represent different stages of the process. Values are means ± SE.

Such changes in the structure of microbial communities were accompanied by a decrease in the abundance of both Gram-positive and –negative bacterial populations with the depth of layers (Figure 3A,B), i.e. from upper to medium and lower layers; and the abundance of these groups were in the fresh rabbit manure 346 ± 49.0 and 336 ± 63 μg g^{-1} dw for Gram-positive and Gram-negative bacteria, respectively (Figure 3A,B). A similar trend was observed for fungal populations (Figure 3C), reaching an average value of 1.3 ± 0.1 at the end of the process (Figure 3C). These results are in accordance with previous studies based on PLFA profiles, with marked changes in the structure of microbial communities due to decreases in both bacterial and fungal populations throughout the process of vermicomposting [18, 89]. Recently, Fernández-Gómez et al. [94] observed that the structure of fungal communities, assessed by DGGE profiles differed at the stage of maximum earthworm biomass the most, suggesting the existence of a strong gut passage effect on the microbial communities through a continuous-feeding vermicomposting system in the presence of *E. fetida*.

Decreases in microbial activity were also detected with depth of layer (Figure 4A) and, after a maturation period for 250 d, basal respiration values dropped below 100 mg CO_2 kg^{-1} OM h^{-1} (Figure 4A), as previously shown by [18]. Accordingly, a reduction in the dissolved organic carbon content was detected from upper to lower layers (Figure 4B), reaching a value close to 7000 μg g^{-1} dw after 250 d of vermicomposting. In contrast, other authors [17] reported levels of DOC much more lower in a long-term experiment (252 days) with the epigeic earthworm *E. fetida*, with values below 1500 μg g^{-1} dw in the presence of earthworms. Such differences could be due to the composition of the parent material (pig slurry *versus* rabbit manure) and/or to the experimental setup conditions. Unlike compost -a limit value of 4000 mg kg^{-1} is suggested for a stable compost according to [34]- there is still no threshold level of DOC for which vermicompost is to be considered stable.

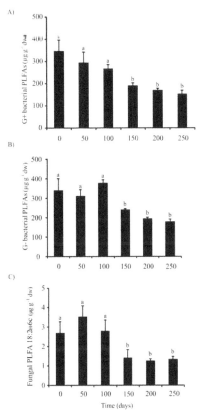

Figure 3. Changes in (a) Gram-positive bacterial, (b) Gram-negative bacterial and (c) fungal PLFAs in the layers of reactors fed with rabbit manure throughout the process of vermicomposting. The different layers represent different stages of the process. Different letters indicate significant differences between the layers based on post hoc test (Tukey HSD). Values are means ± SE.

Overall, in the present study a higher degree of stabilisation was reached in the rabbit manure after a period of between 200 and 250 days, as indicated by the lower values of microbial biomass and activity that are indicative of stabilized materials. These results underscore the potential of epigeic earthworms in the stabilisation of this type of organic substrates, which is of great importance for the application of animal manures as organic amendments into agricultural soils because, as already mentioned, it is widely recognised that the overproduction of this type of substrate has led to inappropriate disposal practices, which may result in severe risks to the environment [6]. Furthermore, these findings constitute a powerful tool for the development of strategies leading to a more efficient process for the disposal and/or management of animal manures, thereby highlighting the continuous-feeding vermicomposting system as an environmentally sound management option for recycling such organic wastes, as previously reported by [94] for treating tomato-fruit waste from greenhouses. Ultimately, it should be borne in mind that the functioning of this type of reactors can lead to the gradual accumulation of layers and compaction of the substrate, thus minimizing earthworm- induced aeration, which can promote pathogen survival [89].

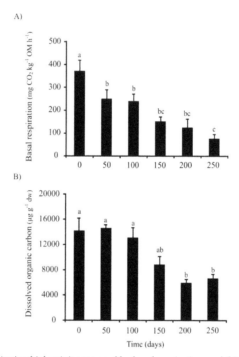

Figure 4. Changes in (a) microbial activity assessed by basal respiration, and (b) dissolved organic carbon content in the layers of reactors fed with rabbit manure throughout the process of vermicomposting. The different layers represent different stages of the process. Different letters indicate significant differences between the layers based on post hoc test (Tukey HSD). Values are means ± SE.

2.2.2. Influence of vermicompost amendments on the soil microbiota

As occurred with compost amendments, vermicompost has also been found to provide manifold benefits when used as a total or partial substitute for mineral fertiliser in peat-based artificial greenhouse potting media and as a soil amendment in field studies [95]. Among the advantages of vermicompost as a soil amendment is its potential to maintain soil organic matter, foster nutrient availability, suppress plant diseases and increase soil microbial abundance and activity. However, although several studies have tried to disentangle the complex interactions between vermicompost application and soil microbial properties, most of them are frequently not comparable to each other due to differences in the experimental design, the land-use and vermicompost type (i.e., different starting material and earthworm species), the dose of application as well as the duration of experiments, among others. Despite these limitations, some recent findings have been made, thereby contributing to better understand whether and to what extent vermicompost amendments affect soil microbial biomass, activity and community structure. For instance, Arancon et al. [96] observed that a single application of vermicompost to a strawberry crop resulted in a significantly higher increase in soil microbial biomass than the application of an inorganic fertiliser, regardless of the dose used. Increases in the microbial activity and in the activity of the soil enzymes involved in the release of the main plant macronutrients with vermicompost amendments, have also been signalled in several studies [96-98]. Such increase could be due to the fact that soil microorganisms degrade organic matter through the production of a variety of extracellular enzymes and, in turn an input of organic matter is expected to be accompanied by a higher enzymatic activity. Moreover, the added material may contain intra- and extracellular enzymes and may also stimulate microbial activity in the soil [99]. Additionally, vermicompost has been found to promote the establishment of a specific microbial community in the rhizosphere different from that of plants supplemented with mineral fertilisers or other types of organic fertilisers such as manure [100]. Inorganic fertilisation only supplies N, P and K, whereas organic fertilisers also supply different amounts of C and macro- and micronutrients, which can select for microbial communities with different nutritional requirements [95]. Moreover, microbial communities in vermicompost are metabolically more diverse than those in manure [17], and may be incorporated, at least in the short-term, to soils [101]. Interestingly, Aira et al. [100] observed that the effect of the addition of vermicompost occurred despite the low dose used (25% of total fertilisation), and despite the short duration of the experiment (four months). Jack et al. [102] also examined how different organic transplant media amendments, including vermicompost, thermogenic compost and industry standard amendments affected the rhizosphere bacterial communities of organically produced tomato plants. They found differences in the bacterial community structure between the different amendments and these differences persisted for at least one month after seedlings were transplanted to the field. Since both compost and vermicompost were made from the same parent material, such differences could be due to the way in which the organic matter was processed prior to the amendment [102]. Previous comparisons between vermicompost and compost with respect to microbial communities [103-105] are difficult to interpret because different feedstocks were used for each process. Compost feedstocks are known to alter the material's effects on the structure of the microbial communities [86], so it is essential to use composts made from the same feedstock in order to draw valid comparisons between the two biological processes. Furthermore, it may be expected that different hybrids or plant genotypes will respond differently to vermicompost, considering that plant genotype determines important differences in nutrient uptake capacity, nutrient use efficiency and resource allocation within the plant. Different genotypes may therefore enhance root growth or modify root exudation patterns in order to increase nutrient uptake [100], and all of these strategies will determine the establishment of different interactions with the microbial communities at the rhizosphere level. Indeed, after the application of vermicompost to sweet corn crops, these authors found important differences in the rhizosphere microbial community of two genotypes from cultivars of maize, with the sugary endosperm mutation (*su1*) and with the shrunken endosperm mutation (*sh2*), which differ in their C storage patterns.

Furthermore, recent studies have demonstrated the presence of various bacteria, which are useful for different biotechnological purposes, in diverse vermicomposts [106-107], reinforcing that the biological component (i.e., the microbial community composition) of a vermicompost largely determines its usefulness in agriculture and other applications, such as soil restoration and bioremediation. For instance, Fernández-Gómez et al. [106] detected the presence of *Sphingobacterium*, *Streptomyces*, Alpha-Proteobacteria, Delta-Proteobacteria and Firmicutes in diverse vermicomposts, irrespective of the parent material used for the process, by applying DGGE and COMPOCHIP, thereby demonstrating the usefulness of both techniques to assess the potential of vermicomposts as bioactive organic materials. Indeed, disease suppressiveness is obviously linked to the microbiota added with the vermicompost, along with the biological and physicochemical characteristics of the native soil microbial community. However, despite the large body of scientific evidence showing the positive effects of vermicompost regarding the suppression of soil-borne plant fungal diseases (reviewed in [75,108]), it is still necessary to obtain a deeper understanding of the mechanisms involved and the main factors influencing such suppressing effects. According to the mechanisms proposed for compost [68-69], disease suppression by vermicompost may be attributed to either direct effects or to the induction of systemic resistance in the plant. Direct suppression of the pathogen by the vermicompost-associated microflora and/or microfauna may be general or specific, depending on the existence of a single suppressive agent or the joint action of several agents, and the proposed mechanisms are competition, antibiosis and parasitism. Some of the indirect effects of vermicompost have been related to changes in the microbiological properties of the soil or the potting media. Processing by earthworms during vermicomposting has a strong effect on the microbial community structure and activity of the initial waste [8]. Vermicompost therefore has a rather different microbial community structure than the parent waste, with lower biomass but enhanced metabolic diversity [18]. Application of such a microbiologically active organic substrate may thus have important effects on the microbial and biochemical properties of soil or greenhouse potting media thereby influencing plant growth. Moreover, vermicompost may affect directly the plant growth via the supply of nutrients, as it constitutes a slow-release fertiliser that supplies the plant with a gradual and constant source of nutrients, and/or through the supply of plant growth regulating substances [95].

3. Anaerobic digestion

The process of anaerobic digestion (AD) has been extensively studied in natural and engineered ecosystems for more than a century. In natural habits, the anaerobic degradation of organic matter takes place in sediments, waterlogged soils and animal intestinal tracts, in which the oxygen access is restricted; whilst in engineered environments it refers to the biotechnological process by which organic matter (i.e., organic waste, wastewater and/or a renewable resource) is degraded in the absence of oxygen for the commercial production of biogas that can be used as an eco-friendly energy source [109], thereby representing an important asset in times of decreasing fossil fuel supplies. According to [4], the anaerobic digestion from swine, bovine and poultry slurries resulted in the production of biogas at average rates of 0.30, 0.25 and 0.48 L/g volatile solids, respectively. Another valuable co-product derived from this process is the anaerobic digested slurry, which can be applied as an organic amendment into soil either in agricultural and non-agricultural lands [110](section 3.1). It is for this reason along with the production of biogas and the reduction in greenhouse gas emissions [10] that anaerobic digestion is becoming increasingly popular as a methodological alternative for manure recycling, which in turn has increased the number of farm-scale anaerobic bioreactors up to 4200 in central and northern Europe [111].

Bacteria represent over 80% of the total diversity in anaerobic digesters [112], and they are mainly composed by the phyla Firmicutes, Proteobacteria and Bacteroidetes; whereas most of the archaeal representatives belong to the phylum Euryarchaeota, which includes all the known methanogens. Anaerobic eukaryotes - particularly, fungi and protozoa- have received less attention probably because they are slower growers than bacteria and, as such, their abundance is lower in anaerobic reactors. As occurs with composting, anaerobic digestion may be described as a four-phase microbiologically driven process. Briefly, the first and rate limiting step of the anaerobic food chain is the *depolymerization* and *hydrolysis* of complex biopolymers, such as polysaccharides, lipids, proteins and nucleic acids into their corresponding structural units (sugars, fatty acids, amino acids, purines and pyrimidines) through the joint action of a complex community of fibrolytic bacteria and fungi, which produce extracellular hydrolitic enzymes (i.e., cellulases, xylanases, proteases, lipases) responsible for the disassembling of such polymers. Since polysaccharides, mainly cellulose, are the most abundant structural and storage compounds of biomass, their hydrolysis is considered as the most determinant enzymatic process regarding the efficiency of anaerobic reactors. The rate and efficiency of cellulose hydrolysis depends greatly on the particular microbial species composition involved [113], and under anaerobic conditions it proceeds slowly due to the heterogeneity of forms in which cellulose is present in nature and to the complexity of the hemicelluloses and lignin matrices in which it is embedded [113]. Similarly, protein hydrolysis to peptides and amino acids takes place slowly, whilst lipid hydrolysis into glycerol and long-chain fatty acids occurs rapidly compared to their subsequent fermentation or β-oxidation. As mentioned above, bacterial populations are more abundant and diverse and, hence, they are responsible for the majority of hydrolytic reactions, being *Clostridium*, *Acetivibrio*, *Bacteroides*, *Selenomonas* or *Ruminococcus* common examples of hydrolytic bacteria found in anaerobic reactors [112,114].

The monomeric compounds released after the hydrolysis of biopolymers can be taken up by microbial cells, in which they are either fermented or anaerobically oxidised into alcohols, short-chain fatty acids, CO_2 and molecular hydrogen (H_2). This step is known as *fermentation* (*acidogenesis*) and usually occurs through the production of an energy-rich intermediate that is used to synthesise ATP, rendering a fermentation product that is excreted out of the cell. Since these products are typically acidic substances, the fermentative reactions are accompanied by a decrease in the extracellular pH. This fact along with the increase in short-chain fatty acids represents the most common reasons for reactor failure. Thus, for maintaining the pH balance of the system, it is of great importance to bear in mind the equilibrium between fermentative-acidogenic bacteria and acid scavenging microbes. Typically, the bacteria from the same group that hydrolyze biopolymers take up and ferment the resulting monomers. For instance, *Clostridium* sp. and enteric bacteria are common sugar fermenters in anaerobic reactors. *Streptococcus* sp. and *Lactobacillus* sp. also ferment sugars, producing lactate or lactate and ethanol, plus CO_2 and molecular hydrogen [114]. Fermentation of amino acids and purines and pyrimidines in anaerobic environments is mainly carried out by species of *Clostridium* [115].

Then, during *acetogenesis*, the fermentation products are mainly oxidised to acetate, formate, CO_2 and H_2 by acetogenic bacteria, most of them belonging to the low G+C branch of *Firmicutes* [116]. Certain acetogenic reactions are thermodynamically unfavourable under standard conditions, which make necessary a syntrophic relationship between the acetogen and a H_2-consuming methanogens in order to degrade the substrate and, in turn to obtain a net energy gain [117]. Finally, the last and most sensitive step during the anaerobic digestion of organic matter is the *methanogenesis*, i.e. the formation of methane from acetate, H_2/CO_2 and methyl compounds by the action of methanogenic organisms belonging to the phylum Euryarchaeota [118]. The orders Methanobacteriales, Methanococcales, Methanomicrobiales and Methanosarcinales include known methanogens commonly found in aerobic reactors. Members of the first three orders use CO_2 and H_2 as an electron acceptor and donor, respectively. Some species from these orders can also use formate and/or secondary alcohols (i.e., isopropanol or ethanol), but they cannot use acetate or C1 compounds such as methanol and methylamines (with the exception of the genus *Methanosphaera* from the order Methanobacteriales). However, Methanosarcinales are more diverse metabolically, and they can use acetate, hydrogen, formate, secondary alcohols and methyl compounds as energy sources. It is believed that the predominance of hydrogenotrophic or acetotrophic methanogens depends on the levels of their substrates and their tolerance to diverse inhibitors, including ammonia, hydrogen sulphide, or volatile fatty acids [119]. The aforementioned steps involved in the anaerobic digestion are explained in more detail by [109].

3.1. Influence of anaerobic digested slurry on the soil microbiota

The changes that occur in the parent waste during the process of anaerobic digestion largely depend on the dynamics of the abovementioned microbial groups, ultimately influencing the quality of the final products, i.e. biogas and anaerobic digested slurry. As mentioned before, anaerobic slurries are rich in partially stable organic carbon and can be used as organic amendments for crop production. However so far, many environmental issues relevant when these co-products are applied to agricultural land still have to be studied, especially those related to the impact of the anaerobic digested slurry on the soil microbiota. In a recent work, Walsh et al. [120] observed that its application affected the fungal and bacterial growth in a very similar way to the application of mineral fertilizers in a 16- week greenhouse experiment. They found a pronounced shift towards a bacterial dominated microbial decomposer community, and such effects were consistent in different soils and different crop types. They conclude that mineral fertiliser could be thus exchanged for anaerobic digested slurry with limited impact on the actively growing soil microbial community, which is of great importance in the regulation of soil processes and consequently in soil fertility and crop yield. Recently, Massé et al. [4] gave an overview of the agronomic value of anaerobic digestion treated manures. In line with this, previous findings showed that the AD of animal slurries improved their fertilizer value [3], thereby leading to an increased forage yield and N uptake relative to raw liquid swine manure and mineral fertilizers [121]. Bougnom et al. [122] found a 20% increase in grass yield compared to conventional manure. Liedl et al. [123] also found that digested poultry litter resulted in similar or superior grass and vegetable yields versus N fertilizers. Therefore, all of the above provides evidence that the anaerobic digested slurry acts similarly to mineral fertiliser and should be considered as such in its application to land. Additionally, the enrichment of mineral fractions of N and P during digestion ultimately results in a higher concentration of plant-available nutrients compared with undigested manure and a subsequently elevated plant growth promotion ability, suggested to be similar to mineral fertilisers [4,123]. These latter authors also found that AD reduced swine manure total and volatile solid concentrations by up to 80% resulting in improved manure homogeneity and lowered viscosity allowing more uniform land application [4]. Nevertheless, the higher levels of mineral N found in the slurry, mainly ammonia, may also lead to an increase in the level of phytotoxicity of the slurry, thereby affecting seed germination and plant growth after land- spreading of this co-product into soils [124]. The presence of other phytotoxic substances, such as volatile fatty acids (i.e., acetic, propionic and butyric acids), as well as the high content of soluble salts may contribute to the slurry phytotoxicity [125]. Furthermore, Goberna et al. [126] found that amending soils with slurry resulted in greater nitrate losses during the first 30 d of a 100 d incubation period in 20 cm-depth lysimeters. In fact, around 23 and 45% of the total N contained in the soil (natural + added) was lost from soils amended with cattle manure and anaerobic slurry, respectively. Other authors also observed that N leaching was, along with NH_3 volatilisation, one of the most important sources of N leakage to the environment in a field-scale experiment, after having quantified the amount of mineral N at 1.7 m depth from grass cultivated plots amended with anaerobic digested slurry and mineral fertiliser [127]. Thus, the use of this co-product as an organic amendment should accurately match crop N demand because if not taken by the plants, nitrates could be drained to surface waters, leached to ground waters or denitrified into gaseous forms and emitted to the atmosphere.

The presence of pathogenic bacteria in agricultural amendments also represents a potential threat and their screening is thus of great importance mainly in those produced from animal manures, as it has been shown that such pathogenic organisms constitute a common fraction of the microbial community in manure [128]. In fact, it has been shown that some pathogenic bacteria can survive the process of anaerobic digestion and persist in the slurry, as previously reported by [129]. In line with this, those microorganisms with a spore-forming capacity such as *Clostridium* and *Bacillus* species, which are commonly found in the intestinal flora of most warm-blooded animals and can harbor some highly pathogenic members for animals and humans [12,129], cannot be reduced during the process [130]. Accordingly, Olsen and Larsen [131] observed that the spores of *Clostridium perfringens* were not inactivated in either mesophilic or thermophilic biogas digesters. Similar results were observed by other authors [132-133] in a reactor operating under mesophilic and thermophilic conditions, respectively. It is acknowledged that bacterial spores can survive in extreme conditions and germinate after long periods, when the conditions become more favourable [131]. The non-hygienic conditions of the storage/transporting tanks can also favour pathogen regrowth [134]. The composition of the substrate fed into the reactor, as well as the reactor conditions such as pH, digestion temperature, slurry hydraulic retention time, ammonium concentration, volatile fatty acids content and nutrient supply are expected to have a significant influence on the sanitation of the end-product [130]. This indicates that there exists a potential risk of spreading potentially pathogenic microbes after the application of anaerobic slurries into soil. Indeed, Crane and Moore [135] stated that amending soils with raw and treated manures, even with a low pathogenic load, still posed a threat for the environment because a period of regrowth of some pathogens including *Escherichia coli*, enterococci, faecal streptococci and *Salmonella enterica* have been shown after manure deposition to soil [136]. Goberna et al. [126] also found that the levels of *Listeria* in soils amended with either cattle manure or anaerobic slurry were significantly higher than those in the control treatment after having been incubated for a month. They observed, however, that the cultivable forms of *Listeria* in the studied soils could correspond to *L. innocua* instead of *L. monocytogenes*, as shown by the polymerase chain reaction assays. However, as recently summarised by [137], anaerobic digestion generally reduces the pathogen risk when compared to untreated substrates.

In a current study, we evaluated, at a microcosm level, the short and long-term effects of the anaerobic digested slurry on soil chemical and microbiological properties compared to its ingestate (i.e., raw manure) and the two widely-recognized products, compost and vermicompost. All of the organic substrates were mixed with soil by turning at a rate of 40 mg N kg^{-1} soil (dry weight). A control treatment that consisted of soil without the addition of any organic amendment was also included. A total of 45 experimental units (5 amendment levels x 3 incubation times x 3 replicates) were set-up in the present study. After an equilibration period of 4 days at 4 °C, 15 columns were dismantled and the sample was collected to analyze (incubation time 0 days). The remaining thirty columns were then maintained in a room at 22 °C, which is the average temperature of the hottest and wettest month in this area and the most suitable for the survival of pathogens. These columns were destructively sampled after 15 and 60 d incubation corresponding to short and medium-term effects. The survival of selected pathogens was then determined according to standard protocols [138-140] (ISO 16649-2, 2001 for *Escherichia coli*; ISO 4832, 1991 for faecal coliforms; and ISO 7937, 2004 for *Clostridium perfringens*) in all the organic materials and amended soils.

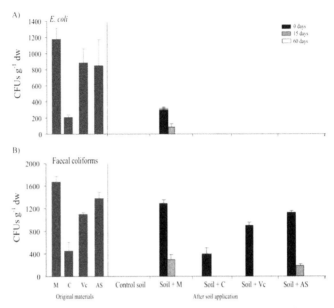

Figure 5. Abundance of *Escherichia coli* and faecal coliforms in the original materials (manure (M), compost (C), vermicompost (Vc) and anaerobic digested slurry (AS)) and in the unamended and amended soils at the three incubation times (0, 15 and 60 days). Values are means ± SE.

Briefly, culturable forms of both faecal coliforms and *E. coli* were isolated from all the initial materials, although their levels were greatly lower in compost relative to the other substrates (Figure 5). This is not surprising taking into account that the composting process, unlike vermicomposting, involved a four-day thermophilic phase, during which the process reached a temperature of 70 °C. Those pathogens were also detected in the anaerobic digested slurry after 40 days of anaerobic digestion (Figure 5). This fact suggests that feeding the reactor with four to five m³ cattle manure d⁻¹ could have provided enough nutrients to maintain a large population of the studied pathogens. Indeed, nutrient availability is one of the major factors influencing pathogen survival in biogas digesters, as previously reported by [130]. Once applied to soils, *E. coli* CFUs were detected in manure-amended soils at the start of the experiment and after incubation for 15 d (Figure 5A); whilst faecal coliforms CFUs were recorded in both manure and slurry-amended soils in the short-term, even though at lower values in comparison with the start of incubation (Figure 5B).

However, the spore-forming *C. perfringens* persisted in all the amended soils (Figure 6), which supports the fact that this bacterium has more resistance to environmental stresses and the capacity to outcompete the native soil microbiota. After 60 d CFU of *C. perfringens* were much closer to those in the control in the slurry-amended soils (Figure 6), which suggests that this time period could be considered as a safe delay between land-spreading the product into soil and crop harvesting with respect to its pathogenic load.

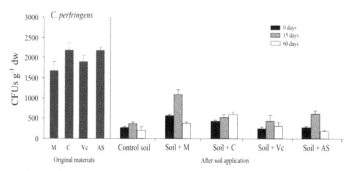

Figure 6. Abundance of *Clostridium perfringens* in the original materials (manure (M), compost (C), vermicompost (Vc) and anaerobic digested slurry (AS)) and in the unamended and amended soils at the three incubation times (0, 15 and 60 days). Values are means ± SE.

4. Conclusions

The intensity and concentrated activity of the livestock industry generate huge amounts of biodegradable wastes, which must be managed with appropriate disposal practices to avoid a negative impact on the environment. Composting is one of the best-known processes for the biological stabilization of solid organic wastes under aerobic conditions. Vermicomposting, i.e. the processing of organic wastes by earthworms under aerobic and mesophilic conditions, has also proven to be a low-cost and rapid technique. Although aerobic processes are thermodynamically more favorable, the manure treatment by anaerobic digestion has become increasingly important due to its energetic potential. A stabilised end product that can be used as an organic amendment is obtained under both aerobic and anaerobic conditions. A multi-parameter approach applying diverse methods constitutes the best option for evaluating the stability/maturity degree of the organic matter, which is of utmost importance for its safe use for the agriculture and the environment. During biodegradation all organic matter goes through the microbial decomposer pool and thus, further knowledge about the changes occurring during the process from a microbial viewpoint will contribute to further develop efficient strategies for the management of animal manures.

5. Outlook

Waste management continues to be a topic of increasing importance. Deeper knowledge of the different biological processes involved in the recycling and recovery of waste components is thus of utmost importance in order to contribute towards more sustainable production and consumption systems. For example, biowaste may be used as a resource to produce high quality lactic acid and protein, as well as biogas in a cascade procedure. Briefly, biowaste is separated into two phases, i.e. a solid phase that is used to feed *Hermetia illucens* larvae that may be harvested as an excellent source of protein for feeding chicken or fish, and the liquid phase that is microbially fermented to the platform chemical lactic acid.

The remaining residuals may eventually be used for biogas production, a cascade process that utilizes the organic waste at its highest level. Furthermore, although an interest in vermicompost research and technology has been increasing over recent years, and the body of knowledge available is quite large, there are still some important topics to be investigated. During vermicomposting, earthworm activity helps microbial communities to use the available energy more efficiently and plays a key role in shaping the structure of the microbial communities during the process. Hence, it is of future interest to evaluate whether the changes in the composition of microbiota in response to earthworm presence are accompanied by a change in the microbial community diversity and/or function. Ultimately, this knowledge will help us to understand the functional importance of earthworms on the stabilization of organic matter from a microbial viewpoint, thereby contributing to minimize the potential risks related to the use of animal manures an organic amendments.

Author details

María Gómez-Brandón*, Marina Fernández-Delgado Juárez and Heribert Insam
University of Innsbruck, Institute of Microbiology, Innsbruck, Austria

Jorge Domínguez
Departamento de Ecoloxía e Bioloxía Animal, Facultade de Bioloxía, Universidade de Vigo, Vigo, Spain

Acknowledgement

Marina Fernández- Delgado Juárez is in receipt of a PhD fellowship from Doktoratsstipendium aus der Nachwuchsförderung der Universität Innsbruck. María Gómez Brandón is financially supported by a postdoctoral research grant from *Fundación Alfonso Martín Escudero*. The authors acknowledge Paul Fraiz for his highly valuable help in language editing.

6. References

[1] Moral R, Paredes C, Bustamante MA, Marhuenda-Egea R, Bernal, MP (2009) Utilisation of Manure Composts by High-Value Crops: Safety and Environmental Challenges. Bioresour. Technol. 100:5454-5460.

[2] Faostat – Food and Agriculture Organization of the United Nations, FAO Statistical Databases (2003). Available: http://faostat.fao.org/.

[3] Holm-Nielsen JB, Seado TAl, Oleskowicz-Popiel P (2009) The Future of Anaerobic Digestion and Biogas Utilization. Bioresour. Technol. 100:5478-5484.

[4] Massé DI, Talbot G, Gilbert Y (2011) On Farm Biogas Production: A Method to Reduce GHG Emissions and Develop more Sustainable Livestock Operations. Anim. Feed Sci. Tech. 166-167:436-445.

---

* Corresponding Author

[5] Steinfeld H, Gerber P, Wasenaar T, Castel V, Rosales M, de Haan C (2006) Livestock's Long Shadow. Environmental Issues and Options. Environmental Issues and Options. Food and Agriculture Organisation (FAO) of United Nations.

[6] Bernal MP, Alburquerque JA, Moral R (2009) Composting of Animal Manures and Chemical Criteria for Compost Maturity Assessment. A Review. Bioresour. Technol. 100: 5444-5453.

[7] Domínguez J, Edwards CA (2010) Relationships between Composting and Vermicomposting: Relative Values of the Products. In: Edwards CA, Arancon NQ, Sherman RL, editors. Vermiculture Technology: Earthworms, Organic Waste and Environmental Management. Boca Raton: CRC Press. pp. 1 - 14.

[8] Domínguez J, Aira M, Gómez-Brandón M (2010) Vermicomposting: Earthworms Enhance the Work of Microbes. In: Insam H, Franke-Whittle I, Goberna M, editors. Microbes at Work. From Wastes to Resources. Berlin Heidelberg: Springer. pp. 93-114.

[9] Darwin C (1881) The Formation of Vegetable Mould through the Action of Worms with Observations on their Habits. Murray, London.

[10] Insam H, Wett B (2008) Control of GHG Emission at the Microbial Community Level. Waste Manage. 28:699-706

[11] Goberna M, Gadermaier M, García C, Wett B, Insam H (2010) Adaptation of the Methanogenic Communities to the Cofermentation of Cattle Excreta and Olive Mill Wastes at 37 °C and 55 °C. Appl. Environ. Microbiol. 101:1899-1903.

[12] Dohrmann AJ, Baumert S, Klingebiel L, Weiland P, Tebbe CC (2011) Bacterial Community Structure in Experimental Methanogenic Bioreactors and Search for Pathogenic Clostridia as Community Members. Appl. Microbiol. Biotechnol. 89:1991-2004.

[13] Wilkinson KG (2011) A comparison of the drivers influencing adoption of on-farm anaerobic digestion in Germany and Australia. Biomass Bioenerg. 35:1613-1622.

[14] Tambone F, Scaglia B, D'Imporzano G, Schievano A, Orzi V, Salati S, Adani F (2010) Assessing Amendment and Fertilizing Properties of Digestates from Anaerobic Digestion through a Comparative Study with Digested Sludge and Compost. Chemosphere 81:577-583.

[15] Liedl BE, Bombardiere J, Chatfield JM (2006) Fertilizer potential of liquid and solid effluent from thermophilic anaerobic digestion of poultry wastes. Water Sci. Technol. 53:69-79.

[16] Aira M, Monroy F, Domínguez J (2006) *Eisenia fetida* (Oligochaeta: Lumbricidae) Activates Fungal Growth, Triggering Cellulose Decomposition during Vermicomposting. Microb. Ecol. 52:738-746.

[17] Aira M, Monroy F, Domínguez J (2007) *Eisenia fetida* (Oligochaeta: Lumbricidae) Modifies the Structure and Physiological Capabilities of Microbial Communities Improving Carbon Mineralization during Vermicomposting of Pig Manure. Microb. Ecol. 54:662–671.

[18] Gómez-Brandón M, Aira M, Lores M, Domínguez J (2011) Changes in Microbial Community Structure and Function during Vermicomposting of Pig Slurry. Bioresour. Technol. 102:4171-4178.

[19] Vivas A, Moreno B, García-Rodríguez S, Benítez E (2009) Assessing the Impact of Composting and Vermicomposting on Bacterial Community Size and Structure, and Microbial Functional Diversity of an Olive-Mill Waste. Bioresour. Technol. 100:1319–1326.

[20] Lazcano C, Gómez-Brandón M, Domínguez J (2008) Comparison of the Effectiveness of Composting and Vermicomposting for the Biological Stabilization of Cattle Manure. Chemosphere 72: 1013-1019.

[21] Fritz JI, Franke-Whittle IH, Haindl S, Insam H, Braun R (2012) Microbiological Community Analysis of Vermicompost Tea and its Influence on the Growth of Vegetables and Cereals. Canadian J. Microbiol., in press.

[22] Jindo K, Suto K, Matsumoto K, García C, Sonoki T, Sanchez-Monedero MA (2012) Chemical and Biochemical Characterisation of Biochar-Blended Composts Prepared from Poultry Manure. Bioresour. Technol. 110:396-404.

[23] Yamamoto N, Asano R, Yoshii H, Otawa K, Nakai Y (2011) Archaeal Community Dynamics and Detection of Ammonia-Oxidizing Archaea during Composting of Cattle Manure Using Culture-Independent DNA Analysis. Environ. Biotechnol. 90:1501-1510.

[24] Gómez-Brandón M, Lores M, Domínguez J (2012) Species-Specific Effects of Epigeic Earthworms on Microbial Community Structure During First Stages of Decomposition of Organic Matter. Plos One e31895.

[25] Zucconi F, de Bertoldi M (1987) Compost Specifications for the Production and Characterization of Compost from Municipal Solid Waste. In: de Bertoldi M, Ferranti MP, L'Hermite P, Zucconi F, editors. Compost: Production, Quality and Use. Barking:Elsevier. pp. 30–50.

[26] Griffin DM (1985) A Comparison of the Roles of Bacteria and Fungi. In: Leadbetter ER, Poindexter JS, editors. Bacteria in Nature. London: Plenum Publishing. pp. 221-255.

[27] Ryckeboer J, Mergaert J, Vaes K, Klammer S, De Clercq D, Coosemans J, Insam H, Swings J (2003) A Survey of Bacteria and Fungi Occurring During Composting and Self-Heating Processes. Ann. Microbiol. 53:349–410.

[28] Beffa T, Blanc M, Lyon PF, Vogt G, Marchiani M, Fischer JL, Aragno M (1996) Isolation of Thermus Strains from Hot Composts (60 to 80 degrees C). Appl. Environ. Microbiol. 62:1723–1727.

[29] Thummes K, Kämpfer P, Jäckel U (2007a) Temporal Change of Composition and Potential Activity of the Thermophilic Archaeal Community During the Composting of Organic Material. Syst. Appl. Microbiol. 30:418-429.

[30] Thummes K, Schäfer J, Kämpfer P, Jäckel U (2007b) Thermophilic Methanogenic Archaea in Compost Material: Occurrence, Persistence and Possible Mechanisms for their Distribution to other Environments. Syst. Appl. Microbiol. 30:634–643.

[31] Vinnerås B, Agostini F, Jönsson H (2010) Sanitation by Composting In: Insam H, Franke-Whittle I, Goberna M, editors. Microbes at Work. From Wastes to Resources. Berlin Heidelberg: Springer. pp. 171-191.

[32] Danon M, Franke-Whittle IH, Insam H, Chen Y, Hadar Y (2008) Molecular Analysis of Bacterial Community Succession during Prolonged Compost Curing. FEMS Microbiol. Ecol. 65:133-144.

[33] Wang P, Changa CM, Watson ME, Dick WA, Chen Y, Hoitink HAJ (2004) Maturity Indices for Composted Dairy and Pig Manures. Soil Biol. Biochem. 36:767-776.

[34] Zmora-Nahum S, Markovitch O, Tarchitzky J, Chen Y (2005) Dissolved Organic Carbon (DOC) as a Parameter of Compost Maturity. Soil Biol. Biochem. 37:2109-2116.

[35] Albrecht R, Joffre R, Gros R, Le Petit J, Terrom G, Périssol C (2008) Efficiency of Near-Infrared Reflectance Spectroscopy to Asses and Predict the Stage of Transformation of Organic Matter in the Composting Process. Bioresour. Technol. 99:448-455.

[36] Tiquia SM (2005) Microbiological Parameters as Indicators of Compost Maturity. J. Appl. Microbiol. 99:816-828.

[37] Wu L, Ma LQ, Martinez GA (2000) Comparison of Methods for Evaluating Stability and Maturity of Biosolids Compost. J. Environ. Qual. 29:424-429.

[38] Cunha Queda AC, Vallini G, Agnolucci M, Coelho CA, Campos L, de Sousa RB (2002) Microbiological and Chemical Characterization of Composts at Different Levels of Maturity, with Evaluation of Phytotoxicity and Enzymatic Activities. In: Insam H, Riddech N, Klammer S, editors. Microbiology of Composting. Berlin Heidelberg: Springer. pp. 345-355.

[39] Insam H, de Bertoldi M (2007) Microbiology of the Composting Process. In: Diaz LF, de Bertoldi M, Bidlingmaier W, editors. Compost Science and Technology. Elsevier Science. pp. 25-48.

[40] Namkoong W, Hwang EY, Cheong JG, Choi JY (1999) A Comparative Evaluation of Maturity Parameters of Food Waste Composting. Compost Sci. Util. 7:55–62.

[41] de Guardia A, Tremier A, Martinez J (2004) Indicators for Determination of Stability Composts and Recycled Organic Wastes. In: Lens P, Hamelers B, Hoitink H, Bidlingmaier W, editors. Resource Recovery and Reuse in Organic Solid Waste Management. London: IWA. pp. 317-337.

[42] Haynes RL, Brinton WF, Evans E (2008) Soil CO_2 respiration: Comparison of Chemical Titration, CO_2 IRGA Analysis and the Solvita Gel System. Renew. Agr. Food Syst.: 23:171–176.

[43] Changa CM, Wang P, Watson ME, Hoitink HAJ, Michel FC Jr (2003) Assessment of the Reliability of a Commercial Maturity Test Kit for Composted Manures. Compost Sci. Util. 11:125-143.

[44] Mondini C, Fornasier F, Sinicco T (2004) Enzymatic Activity as a Parameter for the Characterization of the Composting Process. Soil Biol. Biochem. 36:1587-1594.

[45] Adam G, Duncan H (2001) Development of a Sensitive and Rapid Method for the Measurement of Total Microbial Activity using Fluorescein Diacetate (FDA) in a Range of Soils. Soil Biol. Biochem. 33: 943–951.

[46] Cayuela ML, Mondini C, Sánchez-Monedero MA, Roig A (2008) Chemical Properties and Hydrolytic Enzyme Activities for the Characterization of Two-Phase Olive Mill Wastes Composting. Bioresour. Technol. 99:4255-4262.

[47] Kato K, Miura N, Tabuchi H, Nioh I (2005) Evaluation of Maturity of Poultry Manure Compost by Phospholipid Fatty Acids Analysis. Biol. Fertil. Soils 41:399-410.

[48] Kato K, Miura N (2008) Effect of Matured Compost as a Bulking and Inoculating Agent on the Microbial Community and Maturity of Cattle Manure Compost. Bioresour. Technol. 99:3372-3380.

[49] Evershed RP, Crossman ZM, Bull ID, Mottram H, Dungait JAJ, Maxfield PJ, Emma L, Brennand EL (2006). ^{13}C- labelling of Lipids to Investigate Microbial Communities in the Environment. Curr. Opin. Biotechnol. 17, 72–82.

[50] Lulu B, Egger W, Neunhäuserer C, Caballero B, Insam H (2001). Can Community Level Physiological Profiles be Used for Compost Maturity Testing? Compost Sci. Util. 9: 6-18.

[51] Insam H (1997) A New Set of Substrates Proposed for Community Characterization in Environmental Samples. In: Insam H, Rangger A, editors. Microbial Communities Functional versus Structural Approaches. Berlin Heidelberg: Springer. pp. 260-261.

[52] Insam H, Goberna M (2004) Use of Biolog for the Community Level Physiological Profiles (CLPP) of Environmental Samples. In: Kowalchuk GA, de Brujin FJ, Head IM, Akkermans ADL, van Elsas JD, editors. Molecular Microbial Ecology Manual. Dordrecht: Kluwer. pp. 853-860.

[53] Franke-Whittle IH, Klammer SH, Insam H (2005) Design and Application of an Oligonucleotide Microarray for the Investigation of Compost Microbial Communities. J. Microbiol. Methods 62:37-56.

[54] Franke-Whittle IH, Knapp BA, Fuchs J, Kaufmann R, Insam H (2009) Application of COMPOCHIP Microarray to Investigate the Bacterial Communities of Different Composts. Microb. Ecol. 57:510-521.

[55] Hultman J, Kurola J, Raininsalo A, Kontro M, Romantschuk M (2010) Utility of Molecular Tools for Optimization of Large Scale Composting. In: Insam H, Franke-Whittle IH, Goberna M, editors. Microbes at Work. From Wastes to Resources. Berlin Heidelberg: Springer. pp. 135-152.

[56] Knapp BA, Ros M, Insam H (2010) Do Composts Affect the Soil Microbial Community?. In: Insam H, Franke-Whittle IH, Goberna M, editors. Microbes at Work. From Wastes to Resources. Berlin Heidelberg: Springer. pp. 271-291.

[57] Diacono M, Montemurro F (2010) Long-Term Effects of Organic Amendments on Soil Fertility. A review. Agron. Sustain. Dev. 30:401-422.

[58] Flavel TC, Murphy DV (2005) Carbon and Nitrogen Mineralization Rates after Application of Organic Amendments to Soil. J. Environ. Qual. 35:183-193.

[59] Pérez-Piqueres A, Edel-Hermann V, Alabouvette C, Steinberg C (2006) Response of Soil Microbial Communities to Compost Amendments. Soil Biol. Biochem. 38:460-470.

[60] Innerebner G, Knapp B, Vasara T, Romantschuk M, Insam H (2006) Traceability of Ammonia-Oxidizing Bacteria in Compost-Treated Soils. Soil Biol. Biochem. 38:1092-1100.

[61] Ros M, Klammer S, Knapp BA, Aichberger K, Insam H (2006) Long Term Effects of Soil Compost Amendment on Functional and Structural Diversity and Microbial Activity. Soil Use Manage. 22:209-218.

[62] Saison C, Degrange V, Oliver R, Millar P, Commeaux C, Montange D, Le Roux X (2006) Alteration and Resilance of the Soil Microbial Community Following Compost Amendment: Effects of Compost Level and Compost-Borne Microbial Community. Environ. Microbiol. 8:247-257.

[63] Carrera LM, Buyer JS, Vinyard B, Abdul-Baki AA, Sikora LJ, Teasdale JR (2007) Effects of Cover Crops, Compost, and Manure Amendments on Soil Microbial Community Structure in Tomato Production Systems. Appl. Soil Ecol. 37:247-255.

[64] Kuzyakov Y (2010) Priming effects: Interactions between Living and Dead Organic Matter. Soil Biol. Biochem. 42:1363-1371.

[65] Bending GD, Putland C, Rayns F (2000) Changes in Microbial Community Metabolism and Labile Organic Matter Fractions as Early Indicators of the Impact of Management on Soil Biological Quality. Biol. Fertil. Soils 31:78-84.

[66] Gómez E, Ferreras L, Toresani S (2006) Soil Bacterial Functional Diversity as Influenced by Organic Amendment Application. Bioresour. Technol. 97:1484-1489.

[67] Calbrix R, Barray S, Chabrerie O, Fourrie L, Laval K (2007) Impact of Organic Amendments on the Dynamics of Soil Microbial Biomass and Bacterial Communities in Cultivated Land. Appl. Soil Ecol. 35:511-522.

[68] Hadar Y (2011) Suppressive Compost: When Plant Pathology Met Microbial Ecology. Phytoparasitica 39:311-314.

[69] Noble R, Coventry E (2005) Suppression of Soil-Borne Plant Diseases with Composts: a Review. Biocontrol Sci. Technol. 15: 3–20.

[70] Termorshuizen AJ, van Rijn E, van der Gaag DJ, Alabouvette C, Chen Y, Lagerlöf J et al. (2006). Suppressiveness of 18 Composts Against 7 Pathosystems: Variability in Pathogen Response. Soil Biol. Biochem. 38:2461–2477.

[71] Fuchs JG, Berner A, Mayer J, Smidt E, Schleiss K (2008) Influence of Compmost and Digestates on Plant Growth and Health: Potentials and Limits. In: Paper read at CODIS 2008: Compost and Digestate: Sustainability, Benefits, Impacts for the Environment and for Plant Production. Solothurn, Switzerland.

[72] de Bertoldi M (2010) Production and Utilization of Suppressive Compost: Environmental, Food and Health Benefits. In: Insam H, Franke-Whittle I, Goberna M, editors. Microbes at Work. From Wastes to Resources. Berlin Heidelberg: Springer. pp. 153-170.

[73] Hoitink HAJ, Boehm M J, Hadar Y (1993) Mechanism of suppression of soil borne plant pathogen in compost-amended substrates. In: Hoitink HAJ, Keener HM, editors. Science and Engineering of Composting: Design, Environmental, Microbiological and Utilization Aspects. Worthington, OH, USA: Renais- sance Publications. pp. 601-621.

[74] Nelson EB, Kuter FA, Hoitink HAJ (1990) Effects of Fungal Antagonists and Compost Age on Suppression of *Rhizoctonia* Damping-off in Container Media Amended with Composted Hardwood Bark. Phytopathology 73:1457-1462.

[75] Domínguez J (2004) State of the Art and New Perspectives on Vermicomposting Research. In: Edwards CA, editors. Earthworm Ecology. Boca Raton: CRC Press. pp 401–424

[76] Domínguez J, Edwards CA (2010) Biology and Ecology of Earthworm Species Used for Vermicomposting. Vermiculture Technology: Earthworms, Organic Waste and Environmental Management. In: Edwards CA, Arancon NQ, Sherman RL, editors. Boca Raton: CRC Press. pp 25–37.

[77] Sampedro L, Domínguez J (2008) Stable Isotope Natural Abundances ($\delta^{13}C$ and $\delta^{15}N$) of the Earthworm *Eisenia fetida* and other Soil Fauna Living in Two Different Vermicomposting Environments. Appl. Soil Ecol. 38:91–99.

[78] Monroy F, Aira M, Domínguez J (2009) Reduction of Total Coliform Numbers during Vermicomposting is Caused by Short-Term Direct Effects of Earthworms on

Microorganisms and Depends on the Dose of Application of Pig Slurry. Sci. Total Environ. 407:5411-5416.

[79] Gómez-Brandón M, Aira M, Lores M, Domínguez J (2011) Epigeic Earthworms Exert a Bottleneck Effect on Microbial Communities through Gut Associated Processes. Plos One 6:e24786.

[80] Aira M, Domínguez J (2011) Earthworm Effects Without Earthworms: Inoculation of Raw Organic Matter with Worm-Worked Substrates Alters Microbial Community Functioning. Plos One 6:e16354.

[81] Knapp BA, Podmirseg SM, Seeber J, Meyer E, Insam H (2009) Diet-Related Composition of the Gut Microbiota of *Lumbricus rubellus* as Revealed by a Molecular Fingerprinting Technique and Cloning. Soil Biol. Biochem. 41: 2299-2307.

[82] Aira M, Sampedro L, Monroy F, Domínguez J (2008) Detritivorous Earthworms Directly Modify the Structure, thus Altering the Functioning of a Microdecomposer Food Web. Soil Biol. Biochem. 40: 2511-2516.

[83] Moore JC, Berlow EL, Coleman DC, de Ruiter PC, Dong Q, Johnson NC, McCann KS, Melville K, Morin PJ, Nadelhoffer K, Rosemond AD, Post DM, Sabo JL, Scow KM, Vanni MJ, Wall DH (2004) Detritus, Trophic Dynamics and Biodiversity. Ecol. Lett. 7:584–600.

[84] Schönholzer F, Dittmar H, Zeyer J (1999) Origins and Fate of Fungi and Bacteria in the Gut of *Lumbricus terrestris* L. Studied by Image Analysis. FEMS Microbiol. Ecol. 28:235–248.

[85] Sen B, Chandra TS (2009) Do Earthworms Affect Dynamics of Functional Response and Genetic Structure of Microbial Community in a Lab-Scale Composting System? Bioresour. Technol. 100:804-811.

[86] Lores M, Gómez-Brandón M, Pérez D, Domínguez J (2006) Using FAME profiles for the Characterization of Animal Wastes and Vermicomposts. Soil Biol. Biochem. 38:2993-2996.

[87] Monroy F, Aira M, Domínguez J (2008) Changes in Density of Nematodes, Protozoa and Total Coliforms after Transit through the Gut of Four Epigeic Earthworms (Oligochaeta). Appl. Soil Ecol. 39: 127-132.

[88] Parthasarathi K, Ranganathan LS, Anandi V, Zeyer J (2007) Diversity of Microflora in the Gut and Casts of Tropical Composting Earthworms Reared on Different Substrates. J. Environ. Biol. 28:87-97.

[89] Aira M, Gómez-Brandón M, González-Porto P, Domínguez J (2011) Selective Reduction of the Pathogenic Load of Cow Manure in an Industrial-Scale Continuous-Feeding Vermirreactor. Bioresour. Tehcnol. 102:9633-9637.

[90] Riggle D (1996) Worm Treatment Produces "Class A" Biosolids. BioCycle, 67–68.

[91] Eastman BR, Kane PN, Edwards CA, Trytek L, Gunadi B (2001) The Effectiveness of Vermiculture in Human Pathogen Reduction for USEPA Class A Stabilization. Compost Sci. Util. 9:38-49.

[92] Bardgett RD, Wardle DA (2010) Aboveground-Belowground Linkages: Biotic Interactions, Ecosystems Processes, and Global Change. Oxford: Oxford University Press.

[93] Zelles L (1997) Phospholipid Fatty Acid Profiles in Selected Members for Soil Microbial Communities. Chemosphere 35: 275-294.

[94] Fernández-Gómez MJ, Nogales R, Insam H, Romero E, Goberna M (2010) Continuous-Feeding Vermicomposting as a Recycling Management Method to Revalue Tomato-Fruit Wastes from Greenhouse Crops. Waste Manage. 30:2461-2468.

[95] Lazcano C, Domínguez J (2011) The Use of Vermicompost in Sustainable Agriculture: Impact on Plant Growth and Soil Fertility. In: Miransari M, editors. Soil Nutrients. New York: Nova Science Publishers. pp. 230-254.

[96] Arancon, NQ, Edwards CA, Bierman P (2006) Influences of Vermicomposts on Field Strawberries: Part 2. Effects on Soil Microbiological and Chemical Properties. Bioresour. Technol. 97: 831-840.

[97] Ferreras L, Gomez E, Toresani S, Firpo I, Rotondo R (2006) Effect of Organic Amendments on some Physical, Chemical and Biological Properties in a Horticultural Soil. Bioresour. Technol. 97:635-640.

[98] Saha S, Mina BL, Gopinath KA, Kundu S, Gupta HS (2000) Relative Changes in Phosphatase Activities as Influenced by Source and Application Rate of Organic Composts in Field Crops. Bioresour. Technol. 99:1750-1757.

[99] Goyal S, Chander K, Mundra MC, Kapoor KK (1999) Influence of Inorganic Fertilizers and Organic Amendments on Soil Organic Matter and Soil Microbial Properties under Tropical Conditions. Biol. Fertil. Soils 29:196–200.

[100] Aira M, Gómez-Brandón M, Lazcano C, Bååth E, Domínguez J (2010) Plant Genotype Strongly Modifies the Structure and Growth of Maize Rhizosphere Microbial Communities. Soil Biol. Biochem. 42:2276-2281.

[101] Gómez E, Ferreras L, Toresani S (2006) Soil Bacterial Functional Diversity as Influenced by Organic Amendment Application. Bioresour. Technol. 97:1484-1489.

[102] Jack ALH, Rangarajan A, Culman SW, Sooksa-Nguan T, Thies JE (2011) Choice of Organic Amendments in Tomato Transplants Has Lasting Effects on Bacterial Rhizosphere Communities and Crop Performance in the Field. Appl. Soil Ecol. 48:94-101.

[103] Fracchia L, Dohrmann AB, Martinotti MG, Tebbe CC (2006) Bacterial Diversity in a Finished Compost and Vermicompost: Differences Revealed by Cultivation-Independent Analyses of PCR-Amplified 16S rRNA Genes. Appl. Microbiol. Biotechnol. 71:942–952.

[104] Chaoui HI, Zibilske LM, Ohno T (2003) Effects of Earthworm Casts and Compost on Soil Microbial Activity and Plant Nutrient Availability. Soil Biol. Biochem. 35:295-302.

[105] Anastasi A, Varese GC, Filipello Marchisio V (2005) Isolation and Identification of Fungal Communities in Compost and Vermicompost. Mycologia 97:33-44.

[106] Fernández-Gómez MJ, Nogales R, Insam H, Romero E, Goberna M (2012) Use of DGGE and COMPOCHIP for Investigating Bacterial Communities of Various Vermicomposts Produced from Different Wastes Under Dissimilar Conditions. Sci. Tot. Environ. 414:664-671.

[107] Yasir M, Aslam Z, Kim SW, Lee S-W, Jeon CO, Chung YR (2009) Bacterial Community Composition and Chitinase Gene Diversity of Vermicompost with Antifungal Activity. Bioresour. Technol. 100:4396-4403.

[108] Meghvansi MK, Singh L, Srivastava RB, Varma A (2011) Assesing the Role of Earthworms in Biocontrol of Soil-Borne of Plant Fungal Diseases. In: Karaca A, editors. Biology of Earthworms. Berlin Heidelberg: Springer. pp. 173-189.

[109] Insam H, Franke-Whittle I, Goberna M (2010) Microbes in Aerobic and Anaerobic Waste Treatment. In: Insam H, Franke-Whittle I, Goberna M, editors. Microbes at Work: From Wastes to Resources. Berlin Heidelberg: Springer. pp. 1-34.

[110] Teglia C, Tremier A, Martel JL (2011) Characterization of Solid Digestates. Part 1. Review of Existing Indicators to Assess Solid Digestates Agricultural Use. Waste Biomass Valor. 2: 43-58.

[111] Tabajdi CS (2007) Draft report on Sustainable Agriculture and Biogas: a need for review of EU-legislation (2007/2107 INI), Committee on Agriculture and Rural Development, European Parliament, Brussels.

[112] Krause L, Diaz NN, Edwards RA, Gartemann K-H, Krömeke H, Neuweger H, Pohler A, Runte KJ, Schlnter A, Stoye J, Szczepanowski R, Tauch A, Goesmann A (2008) Taxonomic Composition and Gene Content of a Methane-Producing Microbial Community Isolated from a Biogas Reactor: Genome Research in the Light of Ultrafast Sequencing Technologies. J. Biotechnol. 136:91–101.

[113] Lynd LR, Weimer PJ, van Zyl WH, Pretorius IS (2002) Microbial Cellulose Utilization: Fundamentals and Biotechnology. Microbiol. Mol. Biol. Rev. 66:507–577.

[114] Ueno Y, Haruta S, Ishii M, Igarashi Y (2001) Changes in Product Formation and Bacterial Community by Dilution Rate on Carbohydrate Fermentation by Methanogenic Microflora in Continuous Flow Stirred Tank Reactor. Appl. Microbiol. Biotechnol. 57:65–73.

[115] Schmitz RA, Daniel R, Deppenmeier U, Gottschalk G (2006) The Anaerobic Way of Life. In: Dworkin M, Falkow S, Rosenberg E, Schleifer K-H, Stackebrandt E, editors. The Prokaryotes. A handbook on the Biology of Bacteria. New York: Springer-Verlag. pp 86–101.

[116] Drake HL, Küsel K, Matthies C (2006) Acetogenic Prokaryotes. In: Dworkin M, Falkow S, Rosenberg E, Schleifer K-H, Stackebrandt E, editors. The Prokaryotes. A Handbook on the Biology of Bacteria. New York:Springer-Verlag. pp. 354–420.

[117] Schink B (1997) Energetics of Syntrophic Cooperation in Methanogenic Degradation. Microbiol. Mol. Biol. Rev. 61:262–280.

[118] Briones A, Raskin L (2003) Diversity and Dynamics of Microbial Communities in Engineered Environments and their Implications for Process Stability. Curr. Opin. Biotechnol. 14:270–276.

[119] Demirel B, Scherer P (2008) The Roles of Scetotrophic and Hydrogenotrophic Methanogens during Anarobic Conversion Biomass to Methane: a Review. Rev. Environ. Sci. Biotechnol. 7:173–190.

[120] Walsh JJ, Rousk J, Edward-Jones G, Jones DL, Williams AP (2012) Fungal and Bacterial Growth Following the Application of Slurry and Anaerobic Digestate of Livestock Manure to Temperate Pasture Soils. Biol. Fertil. Soils, doi: 10.1007/s00374-012-0681-6.

[121] Chantigny MH, Angers DA, Rochette P, Belanger G, Massé DI, Côté D (2007) Gaseous Nitrogen Emissions and Forage Nitrogen Uptake on Soils Fertilized with Raw and Treated Swine Manure. J. Environ. Qual. 36:1864–1872.

[122] Bougnom BP., Niederkofler C, Knapp B, Stimpfl E, Insam H (2012) Residues from Renewable Rnergy Production: Their Value for Fertilizing Pastures. Biomass Bionerg. 39:290-295.

[123] Liedl BE, Bombardiere J, Chatfield JM (2006) Fertilizer Potential of Liquid and Solid Effluent from Thermophilic Anaerobic Digestion of Poultry Wastes. Water Sci. Technol. 53:69–79.

[124] Engeli H, EdelmannW FJ, Rottermann K (1993) Survival of Plant – Pathogens and Weed Seeds during Anaerobic-Digestion. Water Sci. Technol. 27:69–76.

[125] McLachlan KL, Chong C, Voroney RP, Liu HW, Holbein BE (2004) Assesing the Potential Phytotoxicity of Digestates during Processing of Municipal Solid Waste by Anaerobic Digestion: Comparison to Aerobic Composts. Acta Hort. 638: 225-230.

[126] Goberna M, Podmirseg SM, Waldhuber S, Knapp BA, García C, Insam H (2011) Pathogenic Bacteria and Mineral N in Soils Following the Land Spreading of Biogas Digestates and Fresh Manure. Appl. Soil Ecol. 49:18-25.

[127] Matsunaka T, Sawamoto T, Ishimura H, Takamura K, Takekawa A (2006) Efficient Use of Digested Cattle Sslurry from Biogas Plant with respect to Nitrogen Recycling in Grassland. International Congress Series 1293: 242-250.

[128] Sidhu JPS, Toze SG (2009) Human Pathogens and their Indicators in Biosolids: a Literature Review. Environ. International 35:187–201.

[129] Bagge E, Sahlström L, Albihn A (2005) The Effect of Hygienic Treatment on the Microbial Flora of Biowaste at Biogas Plants. Water Res. 39: 4879-4880.

[130] Sählström L (2003) A Review of Survival of Pathogenic Bacteria in Organic Waste Used in Biogas Plants. Bioresour. Technol. 87:161-166.

[131] Olsen JE, Larsen HE (1987) Bacterial Decimation Times in Anaerobic Digestions of Animal Slurries. Biol. Waste. 21:153-160.

[132] Chauret C, Springthorpe S, Sattar S (1999) Fate of Cryptosporidium oocysts, Giardia cysts, and Microbial Indicators during Waste- Water Treatment and Anaerobic Sludge Digestion. Can. J. Microbiol. 45:257–262.

[133] Aitken MD, Walters GW, Crunk PL, Willis JL, Farrell JB, Schafer PL, Arnett C, Turner BG (2005) Laboratory Evaluation of Thermophilic-Anaerobic Digestion to Produce Class A Biosolids. 1.Stabilization Performance of a Continuous-Flow Reactor at Low Residence Time. Water Environ. Res. 77: 3019–3027.

[134] Pepper IL, Brooks JP, Gerba CP (2006) Pathogens in Biosolids. Advance. Agron. 90:1-41.

[135] Crane SR, Moore JA (1986) Modeling Enteric Bacterial Die-Off—a Review. Water Air Soil Poll. 27: 411-439.

[136] Sinton L W, Braithwaite R R, Hall CH, Mackenzie ML (2007) Survival of Indicator and Pathogenic Bacteria in Bovine Feces on Pasture. Appl. Environ. Microbiol. 73:7917–7925.

[137] Franke-Whittle IH, Insam H. Pathogen Survival After the Composting, Anaerobic digestion and Alkaline Hydrolysis of Slaughterhouse Wastes: A Review. Critical Reviews in Microbiology (under review)

[138] ISO 4832 (1991) Microbiology-General guidance for the Enumeration of Coliforms-Colony Count Technique, ISO, Geneva.

[139] ISO 16649-2 (2001) Microbiology of Food and Animal Feeding Stuffs-Horizontal Method for the Enumeration of β-glucuronidase-positive Escherichia coli-Part 2: Colony-count Technique, ISO, Geneva.

[140] ISO 7937 (2004) Microbiology of Food and Animal Feeding Stuffs-Horizontal Method for the Enumeration of Clostridium perfringens-Part 2: Colony-count Technique, ISO, Geneva.

Chapter 3

Removal of Carbon and Nitrogen Compounds in Hybrid Bioreactors

Małgorzata Makowska, Marcin Spychała and Robert Mazur

Additional information is available at the end of the chapter

1. Introduction

Biological wastewater treatment methods allow to remove pollutants at high efficiency but they require application of modern knowledge and technology. In bioreactors used for carbon and nutrients removal from wastewater two forms of biomass are utilized: a suspended biomass (dispersed flocs) and an attached biomass (biofilm). The latter needs a carrier on surface which it can grow.

Both types of biomass, despite some similarities, show also many differences. Probably as a result of complex relations (competition, migration, physical factors like flow velocity and biochemical factors like oxygen supply) the flocs and attached biomass can demonstrate many differences, e.g. texture, active surface, heterotrophs and autotrophs ratio, and especially biomass age. A compilation of these two technologies in one hybrid reactor allows to utilize advantages of these technologies and to achieve high carbon and nitrogen removal efficiency. The additional advantages of this new technology (moving bed biological reactor – MBBR; other similar terms: Integrated Fixed Film/Activated Sludge - IFAS, Mixed-Culture Biofilm - MCB) are cost savings and reactor volume reduction. Simultaneous processes maintenance (SND reactor) and specific parameters preservation enable treatment of specific wastewater.

2. Bioreactors' characteristics

2.1. Suspended biomass reactors

Reactors with suspended biomass (activated sludge), commonly used in wastewater treatment, utilize a biocenose of various heterotrophs and autotrophs which are able in certain conditions to remove efficiently pollutants from wastewater. They use dissolved and suspended matter (after hydrolyzing) for biosynthesis and assimilation. One of the basic activated sludge process parameter is the biomass age. This parameter indicates the time of biomass retention in the system and is calculated from biomass balance. The biomass age (sludge retention time, SRT) has an impact on the substrates removal and can be maintained using recirculation, independently on the hydraulic retention time (HRT). On the other hand the pollutions' loads on biomass have a direct impact on the nitrogen and phosphorus removal.

The basic kinetic equation of substrate removal, used in many mathematical models of biological wastewater treatment, is that of Monod type: it describes the substrate utilization rate as a function of specific rate of microorganisms growth [1]:

$$\frac{dS}{dt} = \frac{\mu_{max}}{Y_{max}} \cdot \frac{S}{S+K_S} \cdot X \tag{1}$$

where:
μ_{max} – maximum growth rate, 1/d,
Y_{max} – substrate utilization yield, g_{sm}/g_{sub},
K_S – saturation coefficient, g/m^3,
S – substrate concentration, g/m^3.

In the activated sludge technology three types of reactors are used: continuous stirred tank reactor (fully mixed flow reactor), plug flow reactor and sequencing batch reactor (SBR). The other differentiating factor is aeration of fluid in the reactor, so there are in general three types of reactors: aerated, non-aerated, and intermittently aerated.

Activated sludge flocs have an irregular structure. The disperse rate is related to accessibility of substrate and oxygen for the inert layers of flocs. Li and Bishop [2] prepared microprofiles of the oxygen and substrates concentration in the floc using clark-type microelectrode (figure 1). Redox potential (ORP) changes and oxygen concentrations indicate conditions inside the floc and concentrations of different nitrogen forms describe nitrification process performance (mainly in the top layers of floc) and denitrification (inert layers of floc). The diameter of flocs was in range from 1.0 to 1.4 mm, and oxygen uptake rate was equal to approximately 1.25 mg O_2/dm^3 min.

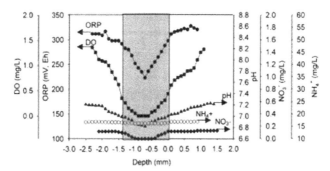

Figure 1. Microprofiles of dissolved oxygen redox potential, pH, nitrates and ammonium concentrations in a floc of activated sludge [2]

2.2. Bioreactors with attached biomass

Attached biomass reactors operate as moving beds, packed (fixed) beds and membranes. The attached biomass (biofilm) has a thickness up to 1.5 mm. The substratum can be fixed (trickling filters and submerged beds) or moving (moving bed biofilm reactors). A main factor affecting access of biomass to the substrate is the effective surface area. The biofilm volume concentration can be even 10 times higher than concentration of activated sludge floc biomass and commonly is in range of 10 to 60 kg/m^3. The biomass age is much longer in biofilm than in flocs and ranges from several to even more than 100 days. The other important factors are: organic substrate loading on substratum surface area and (what is often correlated) the organic substrate loading of the biomass.

The substrate utilization rate in biofilm can be expressed by equation:

$$\frac{dS}{dt} = \frac{\mu_{max}}{Y_{max}} \cdot \frac{S}{K_S + S} \cdot d \cdot X_b \cdot A_b \tag{2}$$

where: d – biofilm thickness, m,
X_b – biofilm density, g/m^3,
A_b – biofilm specific surface area, m^2/m^3.

The substrate penetration into the biofilm depth is affected by SUR (substrate utilization rate) and diffusion coefficient. The relative depth of substrate penetration into the biofilm can be expressed by penetration coefficient (β) assuming that the rate of reaction is zero-order [1]:

$$\beta = \sqrt{\frac{2 \cdot D \cdot S}{k_0 \cdot z^2}} \tag{3}$$

where: D – substrate diffusion coefficient, m^2/s,
k_0 – zero-order reaction constant for biofilm, kg/m^{3}s,
z – biofilm thickness, m.

When β > 1, the substrate penetrates trough the whole depth of biofilm, when β < 1 – substrate penetrates the biofilm only up to the certain depth. When two substrates are considered e.g. oxygen and organic compounds, one of them can be limiting. If the condition: $\beta_{O2} < \beta_{BZT5} < 1$ is satisfied, the conditions in the biofilm will be anaerobic. Conditions and processes in biofilm can be indicated by microprofiles [3]. A dramatic peak of redox potential (figure 2) indicates a change in oxygen conditions – from aerobic near the biofilm surface – to anaerobic – near the substratum. Nitrogen compounds changes indicate the nitrification process caused by oxygen penetration into the subsurface layers of biofilm depth.

2.3. Hybrid bioreactors

A practically useful solution is compilation of the described above two technologies in one reactor named a hybrid reactor. Both activated sludge and biofilm technologies advantages are utilized in this system.

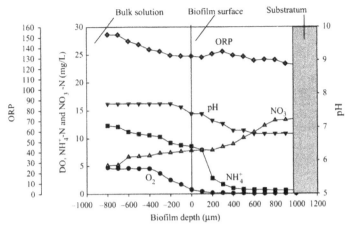

Figure 2. Microprofiles of biofilm [3]

In this type of reactors (Integrated Fixed Film/Activated Sludge - IFAS, Mixed-Culture Biofilm – MCB, hybrid bioreactors) a secondary settler is used and suspended biomass is returned to the bioreactor, so certain suspended biomass concentration can be maintained. However, when suspended biomass flocs are relatively large (up to 1500 μm of diameter) they can clog the small pores of carriers [4] resulting in attached biomass growth interruptions and oxygen access limitations.

Additional modifications can reduce energy consumption. The biofilm substratum may consist of various plastic carriers with effective surface area up to several hundred square meters per cubic meter. The volumetric density of carriers with biomass in fluidized beds should be similar to the wastewater density or slightly higher. There are many market-available types of carriers.

Hybrid reactors with moving carriers were firstly developed in Norway in nineties of XXth century. Characteristics of this technology were given firstly by Odegaard et al. [6]. They proposed the carriers filling rate of 70% of volume and obtained the removal efficiency of 91-94% for organic compounds and of 73-85% for nitrogen compounds. The impact of the substrate loading of reactor on the treatment performance was studied by Orantes and Gonzales-Martinez [7] and Andreottola et al. [8]. These researchers applied this technology to the specific conditions – for resorts in Alps. Andreottola et al. [9] and Daude and Stephenson [10] designed such reactor as a small WWTP for 85 p.e.

The hybrid reactor can be designed basing on organic loading of biomass and knowing the geometry of carriers. The number of carriers (N) can be calculated as [11]:

$$N = \frac{L_S}{A_{k1} \cdot A_b^* \cdot G_b} \tag{4}$$

where: L_S – removed organic load, kg/d,

A_{kl} - one carrier effective surface area , m^2,

A_{b^*} - organic loading of biomass, g/g$_{dm}$d,

G_b – biofilm surface density, g$_{dm}$/m^2.

The simultaneous application of activated sludge and moving bed technologies has a positive influence on the nitrification process. Paul et al. [12] found that 90% of autotrophs in hybrid bioreactor is a component of biofilm (autotrophs are 40% of total number of microorganisms). Despite relatively low kinetic constants of autotrophs growth and substrate utilisation rate comparing to the heterotrophs (Y_H = 0.61 g$_{dm}$/g$_{COD}$, Y_A = 0.24 g$_{dm}$/g$_{COD}$, μ_{Hmax} = 4.55 d^{-1}, μ_{Amax} = 0.31 d^{-1}), the high (over 90%) nitrification efficiency in hybrid reactor can be achieved, even in terms of high hydraulic loading rates.

The nitrifying bacteria in the biofilm on the carriers are able to reach the nitrification rate up to 0.8 gN/m^{2}d at 10°C [13] and even up to 1.0 gN/m^{2}d at 15°C [14].

2.4. Spatial and ecological forms of biomass

There is a little research related to the interactions between activated sludge flocs and biofilm e.g. migration of the organisms. These interactions are complex, and both relations: between flocs and biofilm, and between heterotrophs and autotrophs in the biofilm should be considered in modeling [15] and operation. Albizuri et al. [15] assumed that these interactions could act with mediation of colloidal components. It is well know fact that there are many grazing species (e.g. *Ciliata*) which creep on the flocs/biofilm surface or swim near the flocs and biofilm surface. On the other hand there is some number of species existing in deeper layers of biofilm, which probably can not migrate. It is worth to note that the structure of activated sludge flocs is heterogeneous and deep layers of biomass in flocs are anaerobic, what results in different species composition (anaerobic bacteria).

The biological composition of flocs in hybrid bioreactors is similar to typical biological content of activated sludge flocs. Similarly the size of hybrid bioreactor flocs in hybrid reactors is close to typical activated sludge flocs - diameter in the range of 150-500 µm [16].

In hybrid bioreactors various nitrogen removal processes pathways are possible, including autotrophic processes, e.g. anammox [17]. A sufficiently thick layer of biofilm or flocs is needed for complex nitrogen process transformations including denitrification. On the other hand a relatively thin biofilm (due to shearing stress) results in high activity of biomass [18]. Some authors [19,18] found that in sequencing batch biofilm reactors (SBBR) the biofilm is fully penetrated by substrates and electrons acceptors can be released.

Similarly as in case of other attached biomass systems, e.g. trickling filters in MBBR design procedure, surface area loading rate should be the design parameter [5,11]. This approach is based on some typical range of biofilm thickness (in this case surface area can be the indicator of the biomass concentration). From this point of view the size and shape of carriers seem to be less important. The substrate to biomass loading rate is base but not sole design criterion. Important but poorly recognised factors are: access to total biofilm surface area and access to aerobic biofilm surface area. Some authors indicated that carriers of high total surface area thanks to micropores should have some amount of macropores, enabling fluid reach in oxygen contact with deeper inner spaces of carriers, e.g foamed cellulose carriers [20]. The macropores are important for nitrifying biomass, which needs contact with dissolved oxygen. Micropores are often filled completely with biofilm preventing the oxygen penetration. Oxygen access factor is crucial for biofilm thickness, porosity and surface roughness.

The ratio of the suspended to the attached biomass can vary accordingly to many factors and conditions. The amount of attached biomass can reach over 90%. Plattes et al. [21] indicated 93% of biomass in form of biofilm attached to the carrier elements and only 7% of biomass – as suspended in the bulk liquid. Detachment (or sloughing) of biomass is variable in time [5]; probably this phenomenon is similar to sloughing of excess biomass from biofilm growing in others attached biomass systems, e.g. trickling filters. Some authors [22] suggested that in such systems detachment process occurs periodically.

Due to mechanical contact with others carriers and shear stress, the biomass grows mainly on the internal area of carriers, what was reported by several authors [16, 23], excepting carriers having outgrowths on the outside walls surface.

The common forms in typical activated sludge system are aggregated flocs and planktonic free-swimming cells, and bacterial communities are dominated by: *Betaproteobacteria, Alphaproteobacteria, Gammaproteobacteia* and more less frequent: *Bacteroidetes and Firmicutes* [24]. Some authors [24] observed in biofilms in MBBR limited bacterial diversity and *Firmicutes* domination. The research of Biswas and Turner [24] indicated that MBBR communities differ from communities existing in conventional activated sludge reactors. The characteristic feature of MBBR bacteria community was a presence of two distinct communities: suspended biomass with fast-growing aerobic bacteria and biofilm biomass, which was dominated by anaerobic bacteria [24]. In biofilms of WWTP which were studied by these authors the prevailing forms were *Clostridia* (38% of clones) and sulfate-reducing bacteria (*Deltaproteobacteria* members). The another forms were less abundant: *Desulfobacterales* (11-19%), *Syntrophobacterales* (8-10%), *Desulfovibrionales* (0.5-1.5%). The other groups were also observed: *Bacteroidetes, Synergistes, Planctomycetes, Verrucomicrobia* and *Acidobacteria*.

The suspended biomass observed in two MBBR reactors by Biswas and Turner [24] was consisted mainly of aerobic microorganisms: *Alphaproteobacteria* (*Rhizobiales, Rhodobacterales*), Gammaproteobacteria (*Pseudomonadales, Aeromonadales*), Betaproteobacteria (*Burkholderiales, Rhodocyclales*). Majority of *Firmicutes* was represented by *Clostridia* and one MBBR reactor suspended biomass was reach in *Campylobacteraceae* (54% of clones).

The differences in microbial composition can appear not only between biofilm and activated sludge in MBBR reactor but also between MBBR bioreactors themselves. Biswas and Turner [24] observed the biomass, both black with sulfurous odour in one MBBR reactor and grayish-brown without obvious odour - in other MBBR reactor. Some authors indicated that in continuous-flow MBBR in which SND process was established, the microbial community structures of biofilm are related to C/N ratios [25]. In MBBRs the volume concentration ratio of biofilm to the activated sludge flocs cab be even higher: 5-13 [26] than for separated attached biomass and suspended biomass systems. Some important differences between biofilm and flocs features in MBBRs were found by Xiao end Garnczarczyk [26]. They observed 3 - 5 times higher geometric porosity in biofilm than in activated sludge flocs. Biofilm boundary fractal dimension was higher than activated flocs one. These authors observed also some similarities: two different space populations both in biofilm and in flocs were indicated and both attached and suspended biomass shifted some of their structural properties to larger values (thickness, density) with the increased hydraulic loading.

2.5. Carriers material characteristics and impact on attachment conditions and biomass structure

For the MBBR pollutants removal efficiency and biomass concentration the crucial role plays the material of which carriers were made (table 1). The basic features are as follow: material type, specific surface area, shape and size of carriers and other features: porous surface, e.g. polyurethane or not porous material surface e.g. polyethylene [27,28].

Material	Density, g/cm^3	Specific surface area, m^2/m^3	Type of carrier	References
Cellulose (foamed)	-	-	continuous macro-porous, Aquacel, 1-5 mm	[20]
Polyvinylformal (PVF)	-	-	cubic, 3 mm	[20]
PVA-gel beads	-	-	-	[23]
Polyethylene (PE)	-	-	K1, EvU-Perl	[55]
Nonwoven fabric	-	900	-	[56]
Reticulated polyester urethane sponge (foam)	0.028 (volumetric density)	-	S45R, S60R, S90R, Joyce Foam Products	[57]
Polypropylene (PP)	1.001	230-1400	cylindrical rings 4 mm	[58]
Polyethylene (PE)	0.95	230-1400	cylinders 7 mm/10 mm	[59]
Polyethylene (PE)	-	-	Kaldnes, Natrix, Biofilm-Chip	[60]
Polyurethane coated with activated carbon	-	35,000	cubes (1.3 cm), Samsung Engineering Co	[61]

Table 1. Selected types of carriers and material of which carriers were made

2.6. Carbon and nitrogen compounds removal

The basic role in biochemical transformations have three types of processes:

(*i*) hydrolysis (slow decomposition of polymeric substances to easy biodegradable substances, (*ii*) substrates assimilation by microorganisms correspondingly with Monod equation and (*iii*) growth and decay of microorganisms, what can be written in form:

$$\frac{dX}{dt} = \mu_{max} \cdot f(S) \cdot X - K_d \cdot X \tag{5}$$

where: *f(S)* – substrate concentration related function,
      X – biomass concentration, g_{dm}/m^3,
      K_d – biomass decay constant, 1/d.

Organic substrates in the wastewater are in the form of: suspended solids, colloids and soluble matter. In overall form it can be described as $C_{18}H_{19}O_9N$. Due to their different forms (easy degradable, slowly degradable and not biodegradable fractions) they can be oxidized, assimilated or not biologically decomposed.

The organic substrate fractions, relating to the form and decomposition pathways can be identified correspondingly to commonly used methodology [29-31]. Typical wastewater consists 10-27% of soluble easy biodegradable substances (S_S), 1-10% of not biodegradable soluble substances (S_I), 37-60% of slowly biodegradable suspended solids (X_S) and 5-15% of very slowly biodegradable suspended solids (X_I) [31-34].

Easy biodegradable organic substances are an energy sources for denitrifying and phosphorus accumulating bacteria (PAB) and its concentrations have direct impact on the nitrogen and phosphorus removal.

Nitrogen in wastewater appear usually in form of soluble non-organic forms (mainly ammonium nitrogen, seldom nitrites and nitrates), organic soluble (degradable and not degradable) and as the suspended solids (slowly degradable, not degradable and as a biomass). Nitrogen compounds transformations are carried by autotrophic and heterotrophic bacteria and elementary processes need to preserve adequate technological conditions for these microorganisms.

The polymeric substances hydrolysis is catalysed by extracellular proteolytic enzymes to transform into simple monomers, which can be assimilated by microorganisms.

This process and ammonia nitrogen assimilation is related to fraction of nitrogen in biomass (5-12%). The nitrogen assimilation rate depends on the C/N ratio.

Another process of nitrogen transformation is nitrification – oxidation of ammonium nitrogen to nitrites and nitrates by chemolithotrophs: *Nitrosomonas, Nitrosococcus* and *Nitrosospira* in first phase (hydroxylamine is the intermediate product) and *Nitrobacter, Nitrospira, Nitrococcus* in the second phase, what can be described in form [35]:

$$NH_4^+ + 1,5O_2 \rightarrow NO_2^- + H_2O + 2H^+ - 278 kJ / mol$$

$$NO_2^- + 0,5O_2 \rightarrow NO_3^- - 73 kJ / mol$$

Nitrifying bacteria, as autotrophs, take the energy from carbon dioxide and carbonates. The utilisation rate for *Nitrosomonas* is equal to 0.10 g_{dmo}/g_{N-NH_4}, and for *Nitrobacter* – 0.06 g_{dmo}/g_{N-NO_2} [1]. The important process parameters are: oxygen supply (4.6 g $O_2/1_gN-NH_4$), temperature (5 – 30⁰C), sludge/biomass age (more than 6 days is recommended), biomass organic compounds loading (over 0.2 g BZT_5/g_{dm} is recommended) and BOD/N ratio: when the value is more than five – organic compounds removal dominates, when is lower than three the nitrification is a prevailing process. The decrease in alcanity is the result of nitrification (theoretically: 7.14 g $CaCO_3/1$ g $N-NH_4$) and it causes decrease in pH from 7.5 – 8.5 to 6.5.

Nitrates are transformed in the dissimilation reduction process (*Pseudomonas, Achromobacter, Bacillus* and others) to the nitrogen oxides and gaseous nitrogen correspondingly to the path:

$$2NO_3^- \xrightarrow{-4e} 2NO_2^- \xrightarrow{-2e+2H^+} 2NO \xrightarrow{-2e} N_2O \xrightarrow{-2e} N_2$$

In the terms of dissolved oxygen deficiency (anaerobic or anoxic conditions) those organisms use nitrates as H^+ protons acceptors.

Denitrifying bacteria, as the heterotrophs, need organic carbon for their existence. The source of organic carbon can be: organic compounds in wastewater (internal source), easy assimilated external source of organic carbon, e.g. methanol/ethanol, or intracellular compounds as an energetic source. This ratio of organic carbon should be in range of 5 – 10 g COD/g $N-NO_3$ [35].

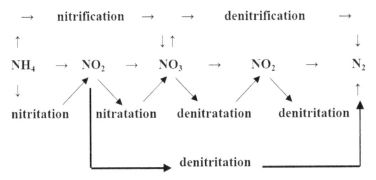

Figure 3. Elementary processes of nitrification and denitrification in nitrogen treatment

In the figure 3 the elementary processes of nitrification and denitrification in nitrogen removal are shown [11]. It is easy to recognise that denitrification is partly the reverse process to the nitrification. Due to the fact that some nitrifying bacteria can live without oxygen and some denitrifying bacteria can survive in oxygen conditions there is the possibility to carry out the simultaneous nitrification and denitrification (SND) in one reactor. As the SND reactor both continuous and sequencing batch reactors can be used. In the SBR-SND reactor high removal efficiency for organic and nitrogen compounds can be achieved – 79% and 96% respectively [36]. The biomass growth rate can be in range of 0.3 – 0.75 g$_{SDM}$/g$_{sub\ rem}$ [37]. Similarly high nitrogen removal efficiency in SND process in continuous reactor (about 90%) can be achieved at certain pH and N-NH$_4$ concentration [38]. The nitrite concentration rise indicates that the nitrification process has stopped after the first phase. The process can run at low C/N ratio.

The short version of SND process needs preservation of certain technological conditions. Important conditions are aerobic and anoxic conditions, what in one reactor system can be achieved by intermittent aeration and non aeration. It causes the characteristic variability of parameters such: pH, redox potential or nitrogen compounds concentration [39]. Examples of changes of some variables in bioreactor with intermittent aeration are presented in figure 4 [40].

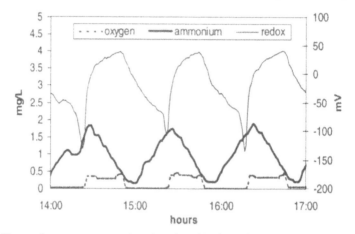

Figure 4. Change of sewage parameters for oxic and aerobic phase of biological reactor [40]

The blockage or limitation of second phase of nitrification can be achieved by limitation of oxygen availability up to approximately 0.7 mg O$_2$/dm^3 [41] or by free ammonia inhibition, which concentration is impacted by the temperature and pH of wastewater. Anthonisen et al. [42] identified the limiting values of partial and full inhibition of second phase of nitrification: 0.1 g N-NH$_4$/m^3 and 1.0 g N-NH$_4$/m^3 respectively. They proposed the description of free ammonia concentration in the reactor in form [42]:

$$S_{NH_3} = \frac{17}{14} \cdot \frac{S_{N-NH_4} \cdot 10^{pH}}{\exp(6344/T) + 10^{pH}}$$

(6)

where: S_{N-NH4} – ammonium nitrogen concentration, mg/dm^3,
         T – temperature, K.

The free ammonia concentration has an impact on the ammonium nitrogen removal velocity, what is described by substrate inhibition model presented by Haldane [43]:

$$r_{NH} = r_{NH\,max} \cdot \frac{S_{NH_3}}{K_{sNH3} + S_{NH_3} + S_{NH_3}^2 / K_{iNH3}} \qquad (7)$$

where: r_{NH} – ammonia nitrogen and ammonium removal velocity, mgN/mg$_{sm}$h,

r_{NHmax} – ammonia nitrogen and ammonium removal maximum velocity, mgN/mg$_{sm}$,

K_{sNH3} – saturation constant, mgN-NH$_3$/dm^3,

K_{iNH3} – inhibition constant, mgN-NH$_3$/dm^3.

For the mathematical description of organic and nitrogen compounds removal many existing models are used with certain modifications, e.g.: substrate inhibition in Brigs–Haldane model [44] or ASM1 model [45], intermittent aeration or oxygen limitation inhibition in ASM1 [46-48], or two stages process of nitrification and denitrification in ASM3 model [49].

3. Laboratory research

The aim of this study was to determine elementary processes related to the organic and nitrogen compounds removal in hybrid reactors with intermittent aeration, to assess removal efficiency under various organic and hydraulic loadings and organic and nitrogen compounds' utilization rates. The utilitarian aim was to determine technological conditions which could make the process shorter and more economically efficient. The attempts to modeling using various technical parameters (together or separately) were conducted.

3.1. Laboratory model and methods

Carriers used in the research were corrugated cylindrical rings made of PP diameter and length of 13 mm, 0.98 g/m^3 density and 0.86 porosity (figure 5).

The research studies were conducted in four stages of 10 months duration. Each stage was consisted of three or four series. In each stage three reactors worked simultaneously as continuous flow system in stage I and III and as a sequencing batch reactor in stage II. The volume of reactors was equal to 75 dm^3 and volume of settler for the continuous flow was equal to 20 dm^3.

The studies were focused on an intermittent aeration. The most attention was put on the last stage – with increased wastewater pH value using lime (Ca(OH)$_2$). The aim of higher pH maintaining was to inhibit the second phase of nitrification by ammonia. The wastewater originated from one family household. The retention time before the sewage discharging into the reactors was relatively high – about 6 days (septic tank and retention tank). The activated sludge originated from Poznań Central WWTP and was inoculated to each reactor at the same amount in each stage beginning. Mixing and aeration of the reactors was made using large-bubble diffusers. The air was supplied by compressor of 0.1-2.0 m^3/h capacity. The sludge recirculation was made using an air-lift cooperating with a membrane pump. Characteristic research parameters are shown in table 2.

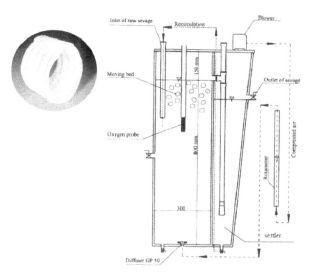

Figure 5. Scheme of laboratory model

	Continuous flow reactor (CFR)	Batch reactor (SBR)	Continuous flow reactor (CFR) with increased pH
Sewage volume per day, dm³/d	70 - 290	45 - 270	140
Number of series	4	3	3
Variable factor for series	time of aeration	length of reactor cycle	number of carriers
Variable factor for reactors	hydraulic load pollution load	active volume of reactor	pH value
C/N	1.31 ± 0.09	1.67 ± 0.10	1.39 ± 0.07
Total solids, g/m³	34.95 ± 6.65	62.30 ± 3.30	37.00 ± 5.20
Organic compounds as COD, g O_2/m³	188.25 ± 1.62	172.80 ± 3.7	184.00 ± 8.00
Nitrogen compounds as N_{tot}, g N/m³	46.56 ± 1.35	41.65 ± 1.09	52.92 ± 2.43

Table 2. Technological characteristics of model investigation and average concentrations of pollutants in sewage

The characteristic feature of the used sewage was a low C/N ratio caused by pretreatment in a septic tank. The pollutants in sewage and suspended biomass concentrations were measured according to the standard methods. The attached biomass concentration was identified via the Kjeldahl nitrogen measurement: 1 g N_{TKN} corresponds to 0.11 g_{dm} [50]; pH and oxygen were measured using calibrated electrodes.

The detailed description of research results of all experimental stages is included in the Makowska's monography [11]. In this chapter only the most important processes and parameters related to the carbon and nitrogen removal efficiency are presented. Results related to the parameters like: biomass loading, pollutants' removal efficiency and substrates utilization rates were analyzed statistically.

3.2. Carbon and nitrogen compounds removal efficiency in continuous and sequencing flow MBBR reactors

The oxygen accessibility play a very important role in bioreactor performance. Four aeration/nonaeration time intervals were tested: 75/45, 45/45, 30/30 and 15/15 minutes. The most effective was the last interval in which oxygen deficit lasted 10 min was observed in nonaeration phase and maximum for SND process oxygen concentration (0.8 mg O_2/dm^3) was achieved. In these conditions for continuous flow, the maximum removal efficiencies for carbon and nitrogen compounds were equal to 98% and 85% respectively [16,11]. The optimal hydraulic retention time was equal to 12 hours.

The outflow pollutants concentrations were related mainly to biomass loading: the higher loading – the lower the removal efficiency, especially for medium and high loaded reactors (the most evident for sequencing batch reactor). This relationship was more evident for nitrogen compounds removal (figure 6) excepting total nitrogen removal in SBR.

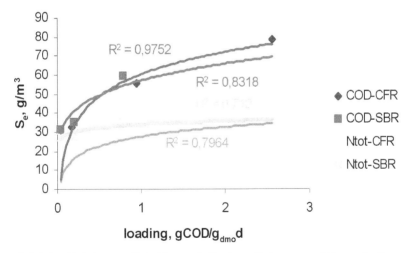

Figure 6. Relationship between pollutants' concentration in purified sewage and biomass loading

The increase in loading up to 2.5 g COD/g_{dmd} caused the rise of contaminants removal rate (figure 7), although it was partly related to biomass concentration decrease as a result of lading rise. The removal efficiency in SBR was related to the volumetric exchange ratio (0.2-0.5 range); the higher ratio – the lower removal efficiency.

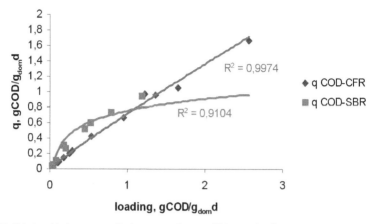

Figure 7. Relationship between pollution removal rate and biomass loading

Some authors stated the higher resistance for hydraulic overloading and more stable nitrification in hybrid reactors than in conventional activated sludge reactors [12]. These research showed that suspended/attached biomass ratio was related to the biomass organic compounds loading (figure 8). The higher biomass loading – the higher attached biomass concentration (the same - lower suspended biomass concentration). This phenomenon was also observed by other authors [26].

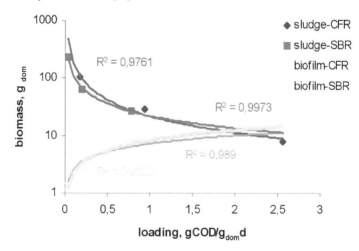

Figure 8. Relationship between biomass in reactors and organic load

So the conclusion can be drown that highly loaded reactors (especially SBR) do not need excess sludge removal, although biomass growth yield can reach values in range 0.31 - 0.50 $g_{dm}/g_{sub\ rem}$.

3.3. Carbon and nitrogen compounds removal efficiency in continuous flow reactor with elevated pH of sewage

Continuous flow reactor with elevated wastewater pH was maintained at 12 hours HRT and 15/15 minutes aeration/nonaeration intervals. Values of pH were in the range of 8.0-8.5. Three rates of volume reactor filling by carriers were investigated: 60%, 40% and 20%.

The elevated pH caused the ammonia release and inhibition of the second phase of nitrification and this way the total nitrogen removal process was shorten by elimination of two elementary processes: nitratation and denitratation (figure 8).

The free ammonia concentration value, calculated accordingly to equation 6 was equal to appr. 1 mg N/dm^3, what is known as a limiting value for inhibition of nitrification second phase [42]. The part of free ammonia could be released as the result of amino acids denaturization during alkalinity process, but due to relatively low concentration of organic nitrogen (mainly amino acids) in inlet wastewater (maximum 10% of total nitrogen) this factor can be neglected. In these conditions the SND process was achieved (shortened nitrogen removal process) what had been indicated by temporary nitrites accumulation. It is known and was stated by several authors [51] that as a result of ammonia inhibition of second phase of nitrification, mediate and final products are released simultaneously.

The higher removal efficiency was achieved at higher volume carriers filling of reactor (figure 9). The rise in pH value versus the rise of nitrogen compounds removal rate, what was observed by other authors [52].

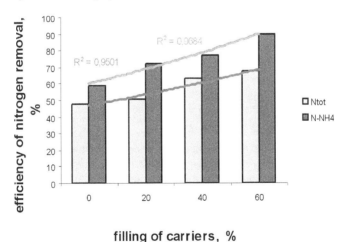

filling of carriers, %

Figure 9. Relationship between effect of nitrogen removal efficiency and filling by carriers

The lime addition resulted in some changes in biomass, e.g. reducing organic fraction rate in biomass of 25% comparing to the process without lime addition. The lime in the liquid phase and on the carriers surface were some kind of "condensation centers" and caused the higher concentration of both biomass form. The smaller amount of carriers enabled more undisturbed carriers movement (figure 10). The lime addition caused also the less susceptibility of carriers pores for clogging by biomass.

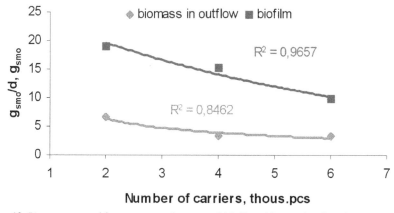

Figure 10. Biomass removed from reactor and amount of biofilm with quantity of carriers

3.4. Hybrid bioreactors biomass characteristic

In this research relatively wide (but typical for activated sludge reactors) range of activated sludge flocs size was observed: 150-500 μm in all reactors.

The biomass attached to the moving bed carriers surfaces was poorly developed. Attached biomass did not cover the outside carriers surface and existed only on the inert surface not as continuous film but as separate small mushroom shape colonies. Small colonies of stalked ciliates (figure 11) were observed on the inner surface of carriers. There were observed some differences between biofilm and activated sludge flocks groups of organisms, in both biomass forms. Stalked, creeping and free-swimming ciliates, filamentous microorganisms, rotifers and nematodes were observed.

Epistylis and *Vorticella* were dominating genera of ciliates in both suspended and attached biomass. Stalked ciliates were observed in relatively high number both in attached and suspended biomass. The domination of this form of *Ciliata* was probably related to good pollutant removal efficiency, what was reported by other authors [1, 54].

The observed number of rotifers in the attached biomass was much higher than in suspended biomass (t-Student statistics; t calculated: 2.83, critical value: 2.78, α: 0.05, replications no.: 5, df: 4), what was the most evident in period 3 in SBRs, and was probably related to the long time of growth of rotifers. Also the differences in concentrations of filamentous microorganisms were observed - in continuous flow reactors the number of filamentous microorganisms was often lower in attached biomass than in suspended biomass (figure 12). Filamentous microorganisms concentration was the highest in the highest COD loaded reactor: R3 (t-Student statistics; t calculated: 8.98, critical value: 2.2, α: 0.05, replications no.: 13, df: 11).

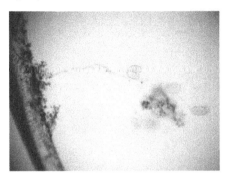

Figure 11. The inner surface of a carrier covered by a small stalked ciliates colony [16]

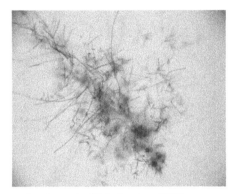

Figure 12. Filamentous organisms in reactor R2 during stage II [16]

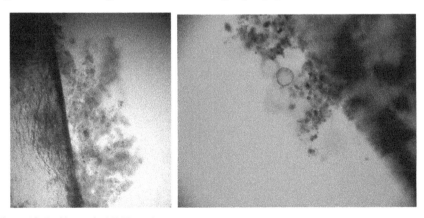

Figure 13. Biofilm on the MBBR carrier

3.5. Hybrid bioreactor mathematical modeling

The mathematical modeling is a very useful method of process simulation, because there is no need time and costs consuming experimental methods using. However, there are many problems in adequate mathematical description of complex biochemical processes and parameters' estimation.

For hybrid bioreactor modeling the ASMH1 model [11] can be applied. It is based on the ASM1 model, but significantly modified: nitrogen removal processes are more completely treated by implementation of two stages nitrification and denitrification (figure 14). The intermittent aeration and oxygen accessibility to the second phase of nitrification was considered. The free ammonia inhibition was also implemented.

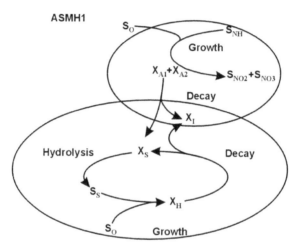

Figure 14. Scheme of process in the ASMH1 model

The model calibration using laboratory data allowed to identify the kinetic and stechiometric parameters values. The model was implemented in to the POLYMATH program and relatively good agreement with experimental results was achieved (st. dev. 10%).

4. Conclusions

The basic technological parameters related to removal efficiency of pollutants from septic tank in hybrid MBBR, at intermittent aeration, continuous/sequencing flow and elevated pH were presented in table 3. The parameters' values concerning the purified sewage fulfill the Polish law requirements for 2000 p.e. WWTPs.

The hybrid MBBR has occurred an effective system for carbon and nitrogen compounds removal from septic tank effluent. The carbon and nitrogen compounds can be removed with at least 80% and 50% removal efficiency respectively. It can be achieved even at loading of 2 g COD/g_{dm}d and 12 hours HRT. Similar results were reported by [53] for partial nitrification-denitrification process in combination of aerobic and anoxic reactors with Kaldnes carriers. The system needed internal recirculation. Thanks to attached biomass the nitrification process in the hybrid MBBR was effective at low and high loaded reactors. The remaining of ammonium nitrogen in treated wastewater appeared at high loadings only. The intermittent aeration and dissolved oxygen limitation enabled simultaneous nitrification-denitrification process (SND) in one reactor. The inhibition of second phase of nitrification by free ammonia has intensified nitrogen removal and resulted in energy savings and internal source of carbon using as a sole source. The shortened process of carbon compounds removal was confirmed by medium products appearance. The long time aeration cycles and long time operating cycles can result in denitrification disturbances due to the organic substances oxidation and limitation of that energy source for denitrifying bacteria.

Parameter	Continuous flow reactor CFR	Batch reactor	CFR with increased pH
Pollution load -organic compounds, g COD/d -nitrogen compounds, g N_{tot}/d	19.29 – 29.25 5.93 – 6.42	15.56 – 31.57 2.77 – 7.87	26.51 – 27.89 7.30 – 8.25
Hydraulic load, dm³/dm³d	1.53 – 1.90	1.65 – 3.05	1.85 – 2.04
Concentration in purified sewage -organic compounds, g COD/m³ -nitrogen compounds, g N_{tot}/m³	26.91 – 56.22 21.57 – 31.07	31.31 – 59.46 27.49 – 34.94	23.00 – 37.00 17.10 – 25.28
Biomass: -activated sludge density, g/dm³ -biofilm mass, g/m²	0.62 – 1.18 2.58 – 5.46	0.48 – 4.03 0.62 – 3.70	5.58 – 6.52 1.55 – 3.58
Biomass loading of -organic compounds, g COD/g_{dm}d -nitrogen compounds, g N_{tot}/g_{dm}d	0.247 – 0.945 0.067 – 0.201	0.081 – 1.080 0.011 – 0.155	0.075 – 0.082 0.011 – 0.021
Efficiency of removal,% -organic compounds -nitrogen compounds	75 – 84 37 – 54	69 – 82 20 – 35	80 – 87 51 - 68
Pollution removal rate -organic compounds, g COD/g_{dm}d -nitrogen compounds, g N_{tot}/g_{dm}d	0.206 – 0.663 0.030 – 0.113	0.110 – 0.620 0.004 – 0.032	0.064 – 0.124 0.011 – 0.021
Yield coefficient, g_{dm}/g_{CODrem}	0.31 – 0.43	0.42 – 0.50	0.34 – 0.47

Table 3. Technological parameters and treatment efficiency in hybrid reactors

The pH elevation brought about higher treatment efficiency. The higher volumetric fraction of moving media (carriers) – the better performance. The continuous flow reactor was more effective in treatment and more stable than the sequencing batch reactor.

It was stated and statistically confirmed that: aeration regime, biomass loading and media volume fraction have an impact on the pollutants (especially organic compounds) removal efficiency.

Advantages of hybrid MBBR reactors operating in the modified conditions are as follows: lower energy consumption (up to 40%) related to the shorter aeration time, possibility of specific wastewater treatment (low N/C ratio), simultaneous processes maintaining in one reactor (internal recirculation elimination), overloading resistance (stable performance) and reduction in smaller reactors' volume. The mathematical model ASMH1 allows to simulate the reactor performance at the specific conditions.

Author details

Małgorzata Makowska, Marcin Spychała and Robert Mazur
Poznan University of Life Sciences, Department of Hydraulic and Sanitary Engineering, Poznań, Poland

Acknowledgement

The authors gratefully acknowledge the funding of this investigation by the Polish State Committee for Scientific Research (grant No. 3 PO6S 07323)

5. References

[1] Henze M., Harremoes P., Jansen J.C., Arvin E. Wastewater Treatment. Biological and Chemical Processes. Springer-Verlag Berlin 2002

[2] Li B., Bishop P.L. Micro-profiles of activated sludge floc determined using microelectrodes. Water Res. 2004;38(5) 1248-1258

[3] Bishop P.L. The role of biofilms in water reclamation and reuse. Water Sci. Tech. 2007;55(1-2) 19-26

[4] Henze M., van Loosdrecht M.C.M., Ekama G.A., Brdjanovic D. Biological Wastewater Treatment. IWA Publishing 2008

[5] Odegaard H., Gisvold B., Striskland J. The influence of carrier size and shape in the moving bed biofilm process. Water Sci. Tech. 2000;41(4-5) 383-391

[6] Odegaard H., Rusten B., Wessman F. State of the art in Europe of the moving bed reactor (MBBR) process. WEFTEC 2004

[7] Orantes J.C., Gonzales-Martinez S. A new low-cost biofilm carrier for the treatment of municipal wastewater in a moving bed reactor, pp. 863-870. V Specialized Conference on Small Water and Wastewater Treatment Systems. Istanbul – Turkey, 24–26 September 2002

[8] Andreottola G., Damiani E., Foladori P., Nardelli P., Ragazzi M. Treatment of mountain refuge wastewater by fixed and moving bed biofilm systems, pp. 313-320. V Specialized Conference on Small Water and Wastewater Treatment Systems. Istanbul – Turkey, 24–26 September 2002

[9] Andreottola G., Foladori P., Gatti G., Nardelli P., Pettena M., Ragazzi M. Upgrading of a small overloaded activated sludge plant using a MBBR system, pp. 743-750. V Specialized Conference on Small Water and Wastewater Treatment Systems. Istanbul – Turkey, 24–26 September 2002

[10] Daude D., Stephenson T. Moving bed biofilm reactors: the small-scale treatment solution pp. 871-888. V Specialized Conference on Small Water and Wastewater Treatment Systems. Istanbul – Turkey, 24–26 September 2002

[11] Makowska M. Simultaneous removal of carbon and nitrogen compounds from domestic sewage in hybrid bioreactors. (in Polish). Rozpr. Nauk. 413. Wyd. UP, Poznań 2010

[12] Paul E., Wolff D.B., Ochoa J.C., da Costa R.H.R.:Recycled and virgin plastic carriers in hybrid reactors for wastewater treatment. Water Environ. Res. 2007;79(7) 765-774

[13] Bengtsson J., Welander T., Christensson M. A pilot study for comparison of different carriers for nitrification in KALDNES™ moving bed biofilm process. Abstract Handbook of IWA Biofilm Technologies Conference, Singapore 2008

[14] Falletti L., Conte L., Milan M. (2009) Nitrogen removal improvement in small wastewater treatment plants with hybrid and tertiary moving bed biofilm reactors. Proceedings of the 2nd IWA Specialized Conference, Nutrient Management in Wastewater Treatment Processes, Kraków, Poland, September 6-9, 803-810, 2009

[15] Albizuri L., van Loosdrecht M.C.M., Larrea L. (2009) Extended Mixed-Culture Biofilms (MCB) Model to Describe Integrated Fixed Film/Activated Sludge (IFAS) Process Behaviour. Proceedings of the 2nd IWA Specialized Conference, Nutrient Management in Wastewater Treatment Processes, Kraków, Poland, September 6-9, 2009

[16] Makowska M., Spychała M., Błażejewski R. Treatment of Septic Tank Effluent In Moving Bed Biological Reactors with Intermittent Aeration. Polish J. Environ. Stud. 2009; 18(6) 1051 – 1057

[17] Żubrowska – Sudoł M.:Nitrogen transformation analysis in a sequential batch reactor with suspended bed. (in Polish) GWiTS 2002;11, 420-426

[18] Żubrowska – Sudoł M. Use of movable deposit in sewage treatment technology. (in Polish) GWiTS;7-8; 7-8, 266-269

[19] Morgenroth E., Obermayer A., Arnold E., Bruehl A., Wagner M., Wilderer P.A.: Effect of long term idle periods on the performance of sequencing batch reactors. IV IAWQ Specialised Conference on Small Wastewater Treatment Plants Stratford - upon – Avon, UK, 18–21 April 1999

[20] Matsumura M., Yamamoto T., Wang P., Shinabe K., Yasuda K. Rapid nitrification with immobilized cell using macro-porous cellulose carrier. Wat. Res. 1997; 31(5) 1027-1034

[21] Plattes M., Henry E., Schosseler P.M., Weidenhaupt A. Modelling and dynamic simulation of a moving bed bioreactor for the treatment of municipal wastewater. Biochem. Eng. J. 2006;32, 61-68

[22] Lewandowski Z., Beyenal H., Myers J., Stookey D. (2007). The effect of detachment on biofilm structure and activity: the oscillating pattern of biofilm accumulation. Water Sci Tech. 2007;55, 429-436

[23] Levstek M., Plazl I. Influence of carrier type on nitrification in the moving-bed biofilm process. Water Sci. Technol. 2009;59(5) 875-882

[24] Biswas K., Turner S.J. Microbial Community Composition and Dynamics of Moving Bed Biofilm Reactor Systems Treating Municipal Sewage. App. Environ. Microbiol. 2012;78(3) 855-864

[25] Fu B., Liao X., Ding L., Ren H. Charcterization of microbial community in an aerobic moving bed biofilm reactor applied for simultaneous nitrification and denitrification. World J. Microb. Biot. 2010;26, 1981-1990

[26] Xiao G.Y., Ganczarczyk J. Structural features of biomass in a hybrid MBBR reactor. Environ. Technol. 2006;27(3) 289-98

[27] Odegaard H., Rusten B., Westrum T. A new moving bed biofilm reactor – applications and results. Water Sci. Technol. 1994;29, 157-165

[28] Rusten B., McRoy M., Proctor R., Siljudalen J.G. The innovative moving bed biofilm reactor/solids contact reaeration process for secondary treatment of municipal wastewater. Water Environ. Res. 1998;70(5) 1083-1089

[29] ATV-DVWK-A 131P Dimensioning of single-stage activated sludge plants. DWA 2000

[30] Pasztor I., Tury P., Pulai J. Chemical oxygen demand fractions of municipal wastewater for modeling of wastewater treatment. Int. J. Environ. Sci. Tech. 2009;6(1) 51-56.

[31] Myszograj S., Sadecka Z. COD Fractions In Mechanical-Biological Sewage Treatment on the Basis of Sewage Treatment Plant in Sulechów (in Polish). Rocznik Ochrony Środowiska, Tom 6, pp. 233-244, Koszalin 2004

[32] Kappeler J., Gujer W. Estimation of kinetic oarameters of heterotrophic biomass under aerobic conditions and characterization of wastewater for activated sludge modeling . Water Sci. Technol. 1992;25(6) 125-139

[33] Ekama G.A., Dold P.L., Marais G.V.R. Procedures vor determining influent COD fractions and the maximum specific growth rate of heterotrophs in activated sludge systems. Water Sci. Technol. 1986;18(6) 91-114

[34] Mąkinia J., Rosenwinkel K.H., Spering V. Longterm simulation of the activated sludge process at the Hanover-Gruemmerward pilot WWTP. Water Res. 2005;39(8) 1489-1502

[35] Klimiuk E., Łebkowska M. Biotechnology in Environmental Protection (in Polish) PWN, Warszawa 2003

[36] Holman J.B., Wareham D.G. COD, ammonia and dissolved oxygen time profiles in the simultaneous nitrification/denitrification process. Biochem. Environ. J. 2003;22, 125-133

[37] Dobrzyńska A., Wojnowska – Baryła I., Bernat K. Karbon removal by activated sludge dunder Fuldy aerobic conditions At different COD/N ratio. Polish J. Environ. Stud. 2004;13(1) 33-40

[38] Surmacz – Górska J., Cichon A., Miksch K.: Nitrogen removal from wastewater with high ammonia concentration via shorter nitrification and nitrification. Water Sci. Technol. 1997;36(10) 73-78

[39] Corominas Tabares L. Control and optimization of an SBR for nitrogen removal: from model calibration to plant operation. Universitat de Girona 2006

[40] Cecil D. Controlling nitrogen removal using redox and ammonium sensors. Water Sci. Technol. 2003;47(11) 109-114

[41] Ruiz G., Jeison D., Chamy R. Nitrification with high nitrite accumulation for the treatment of wastewater with high ammonia concentration. Water Res. 2003;37(6) 1371-1377

[42] Anthonisen A.C., Loehr R.C., Prakasam T.B.S., Srinath E.G. Inhibitionof nitrification by ammonia and nitrous acid. Journal WPCF 1976;48, 835–852

[43] Antileo Ch., Aspe E., Urrutila H., Zaror C., Roeckel M.: Nitrifying biomass acclimation to high ammonia concentration. J. Environ. Eng. 2002;128(4) 367-375

[44] Bartelmus G., Klepacka K., Gąszczak A., Kasperczyk D. Kinetic of biodegradation of vinyl acetate in sequencing batch reactor. Prace Naukowe Instytutu Inżynierii Chemicznej PAN, Gliwice 2006;8, 5-17

[45] Pambrun V., Paul E., Sperandio M. Control and modeling of partial nitrification of effluents with high ammonia conctrations in Sequencing Batch Reactor, p. 597. IWA Specialized Conference Nutrient Management in Wastewater Treatment Process and Recycle Streams. LEMTECH Konsulting, Kraków 2005

[46] Anderson J.S., Kim H., McAvoy T.J., Hao O.J. Control of an alternating aerobic – anoxic activated sludge system. Part 1: development of a linearization – based modeling approach. Control Eng. Pract. 2000;8, 271-278

[47] Contreras E.M., Ruiz F., Bertola N.C. Kinetic Modeling of inhibition of ammonia oxidation by nitrite under low dissolved oxygen conditions. J. Environ. Eng. ASCE 2008;3, 184-190

[48] Fikar, Chachuat B., Latifi M.A. Optimal operation of alternating activated sludge processes. J. Environ. Tech. ASCE no. 3 pp. 417-424, Control Eng. Pract. 2005;13(3), 853-861

[49] Iacopozzi I., Innocenti V.,Marsili-Libelli S. A modified Activated Sludge Model no. 3 (ASM3) with two-step nitrification – denitrification. Environ. Modell. Softw. 2007;22(6) 847-861

[50] Makowska M., Kolanko H. Kinetic of biomass growth in hybrid reactors (in Polish). Roczniki Akademii Rolniczej w Poznaniu, CCCLXV, Melior. Inż. Środ. 2005;26, 257-265

[51] Glass Ch., Silverstein J. Denitryfication kinetics of high nitrate concentration water: pH effect on inhibition and nitrite accumulation. Water Res. 1998; 32(3) 831-839

[52] Villaverde S., Garcia-Encina P.A., FDZ-Polanco F. Influence of pH over nitrifying biofilm activity in submerged biofilters. Water Res. 2000;34(2), 602-610

[53] Zafarzadeh A., Bina B., Nikaeen M., Movahedian Attar H., Hajian Nejad M. Performance of moving bed biofilm reactors for biological nitrogen compounds removal from wastewater by partial nitrification-denitrification process. Iran J. Environ. Healt. 2010;7(4) 353-364

[54] Ciesielski S., Kulikowska D., Kaczowka E., Kowal P. Characterization of Bacterial Structures in Two-Stage Moving-Bed Biofilm Reactor (MBBR) During Nitrification of the Landfill Leachate. J. Microbiol. Biotechn. 2010;20(7) 1140-1151

[55] Trapani D., Mannina G., Torregrossa M., Viviani G. Hybrid moving bed biofilm reactors; a pilot plant experiment Water Sci. Technol. 2008;57, 1539-1546

[56] Yang S., Yang F., Fu Z., Lei R. Comparison between a moving bed membrane bioreactor and a conventional membrane bioreactor on organic carbon and nitrogen removal. Bioresource Technol. 2009;100, 2369-2374

[57] Guo W., Ngo H., Dharmawan F., Palmer C.G. Roles of polyurethane foam in aerobic moving and fixed bed bioreactors. Bioresource Technol. 2010;101, 1435-1439

[58] Sombatsompop K., Visvanathan C., Aim R.B. Evaluation of biofouling phenomenon in suspended and attached growth membrane bioreactor systems. Desalination 2006;201, 138-149

[59] Colic M., Morse W., Lechter A., Hicks J., Holley S., Mattia C. Enabling the Performance of the MBBR Installed to Trezt Meat Processing Wastewater. CWT 2008; 1-19

[60] Jahren S.J., Rintala J.A., Odegaard H. Aerobic moving bed biofilm reactor treating thermomrchanical pulping whitewater under thermopholic conditions. Water Res. 2002; 36, 1067-1075

[61] Lee W.N., Lee I.J., Lee Ch.H. Factors effecting filtration characteristic in membrane coupled moving bed biofilm reactor. Water Res. 2006; 40, 1827-1835

Biological Activated Carbon Treatment Process for Advanced Water and Wastewater Treatment

Pengkang Jin, Xin Jin, Xianbao Wang,
Yongning Feng and Xiaochang C. Wang

Additional information is available at the end of the chapter

1. Introduction

The development of biological activated carbon (BAC) technology is on the basis of activated carbon technology development. Activated carbon which is used as a kind of absorption medium plays an important role in perfecting the conventional treatment process. Furthermore, activated carbon technology becomes one of the most mature and effective processes to remove organic contaminants in water. Removal of the odor in raw water can be regarded as the first attempt of activated carbon which can play a part in water treatment. The first water treatment plant in which granular activated carbon adsorption tank used was built in 1930 in Philadelphia, United States[1]. In the 1960-1970s, developed western countries started to use activated carbon technology in potable water treatment to enhance the removal of organic contaminants. By then, prechlorination was commonly used as the first step of activated carbon treatment. As the inflow of carbon layer contained free chlorine, the growth of microorganism was inhibited and no obvious biological activity showed in the carbon layer.

In order to improve the removal efficiency of refractory organics, especially the removal of precursors of DBPs, ozonation is commonly used in preoxidation before activated carbon process. The process which combines ozonation and activated carbon treatment was firstly put into practice in the year of 1961 at Amstaad Water Plant in Dusseldorf Germany. The successful trial in Dusseldorf soon arose great attentions from the engineering field in Germany as well as the Western Europe[2]. The advantages of microorganisms growing in the activated carbon layer was first affirmed by Parkhrust and his partners in 1967[3,4], this demonstration enabled the lengthening of the GAC's (Granular Activated Carbon) operation life to a great extent and Ozonation-Biological Activated Carbon technology was finally established. Since early 1970s, the study and application of Ozonation-Biological Activated Carbon treatment were conducted in large scales, among which the major ones are as the followings: the application in water plant of Auf dem Weule, Bremen Germany on a half productive scale[5] and the application in Dohne water plant of Muelheim Germany on productive scale[6]. The successful application of Ozonation-Biological Activated Carbon technology in Germany is widely spread and used in neighboring countries, and the treatment itself was perfected gradually. In late 1970s, the treatment was popularized in Germany. In the year of 1976, the United States Environmental Protection Agency (US EPA) legislated that the activated carbon process must be adopted in potable water treatment process in urban areas with a population over 150,000. Among the water plants using activated carbon treatment, the most representative ones are: Lengg Water Plant in Switzerland[7] and Rouen La Chapella Water Plant in France[8,9], see Fig. 1. the flow diagram. The BAC process was firstly proposed in 1978 by G.W.Miller from the US and R.G.Rice from Switzerland[9]. In 1988, the quality requirements for potable water were improved in Japan and during the years 1988-1992, Kanamachi, Asaka, Kunijima and Toyono water treatment plants using the Ozonation-Biological Activated Carbon process were built[10].

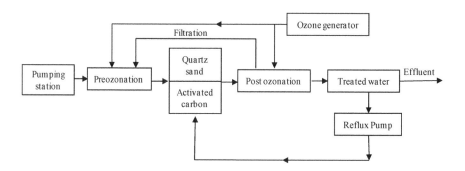

Figure 1. Flow diagram of the water plant in Rouen La Chapella

By now, BAC process has become the major process in advanced water treatment, which is commonly used in developed countries such as America, Japan, Holland, Switzerland, etc[1]. Meanwhile, the process is also widely used in industrial wastewater treatment as well as waste water reclamation. According to the prediction of experts, because of the increasing seriousness of the pollution in potable water and the strictness of requirements for potable water quality, the BAC process which combines the functions of physical-chemical absorption and biological-oxidation degradation, will become the conventional process widely used in potable water treatment plant[9].

2. Composition of biological activated carbon process

2.1. Composition and application

2.1.1. Basic principles of biological activated carbon technology

Biological Activated Carbon process is developed on the basis of activated carbon technology, which uses the synergistic effect of adsorption on activated carbon and biodegradation to purify raw water. Activated carbon has a high specific surface area and a highly developed pore structure, so it is characterized by its great effect on absorbing dissolved oxygen and organics in raw water. For Biological Activated Carbon technology, activated carbon is used as a carrier, by accumulating or artificially immobilizing microorganisms under proper temperature and nutrition conditions, the microorganisms will reproduce on the surface of the activated carbon and finally form BAC, which can exert the adsorption and biodegradable roles simultaneously[11]. The Biological Activated Carbon technology consists of the interaction of activated carbon particles, microorganisms, contaminants and the dissolved oxygen, in water solution. Fig. 2. shows the simplified model that how the 4 factors interact with each other[12]. The relationship between the activated carbon and contaminants is simply the effect of adsorption of activated carbon, and the reaction depends on the properties of the activated carbon and contaminants. Meanwhile, the activated carbon can adsorb DO and microorganisms which were adsorbed on the surface of activated carbon, feed on DO will biodegrade contaminants. In brief, by the interaction of these 4 factors, the purpose for removing contaminant from raw water can be achieved by adopting the biological activated carbon.

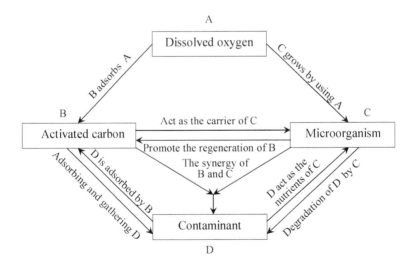

Figure 2. Simplified interaction model of factors in BAC process

2.1.2. Application fields and the typical process flow of biological activated carbon technology

At present, the application of BAC technology is mainly focused on 3 aspects: the advanced treatment for potable water and the industrial waste water treatment. The typical process of the advanced treatment for drinking water and sewage reuse is shown in Fig. 3. All of the three processes are based on conventional coagulation-sedimentation-filtration way. They are distinct from the different positions, the two processing points, to import ozone and the activated carbon. In process a, the activated carbon procedure is between sedimentation and filtration. The outflow from the activated carbon layer will bring some tiny carbon particles and fallen microorganisms, which will be removed by a sand filter in the end. To improve the filtration efficiency, chlorination and enhanced coagulation were done firstly before this procedure. In this process, the quality of the outflow is guaranteed with a relatively higher ozone dosage. While in Process b, the ozonation and the activated carbon procedure is done after the filtration, by which, the ozone depleting substances will be removed therefore a lower ozone dosage than Process a. However, micro carbon particles and microorganisms which leap out of the activated carbon layer will have an undesirable impact on the quality of the outflow, so the frequent backwash on the activated carbon layer is required. Process c is characterized by a two-level ozone procedure, which means to put ozone separately before and after the sand filtration, the remaining procedures are same with Process b. A lower ozone dosage before the sand filtration is used to improve the filtration efficiency[13].

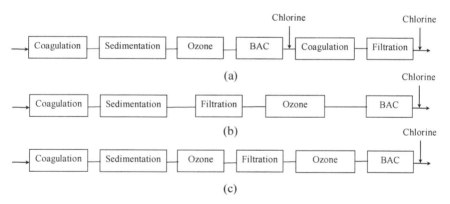

Figure 3. Typical processes of BAC

Also, it is widely used in industrial waste water treatment, such as printing and dyeing wastewater, food processing wastewater, pharmaceutical wastewater, etc. Throughout the typical process of BAC treatment, it is obvious that these three technological processes are related to oxidation-BAC technology. Compared with conventional bio-chemical technology, contact oxidation-BAC process has its unique characteristics. Firstly, contact oxidation can remove organics and ammonia-nitrogen, reduce odor and the amount of DBPs precursor, as well as to reduce the regrowth possibility of bacteria in pipeline, so as to increase the biological stability. Secondly, contact oxidation can reduce the processing load of BAC treatment, and, to some extent, increase the working life and capacity of remaining filtration and BAC, which ensure a safer, reliable outflow[14-17].

2.1.3. Basic operational parameters of BAC process[18]

To design a BAC system, it is necessary to comprehend the characteristics of water quality, water amount and some certain index of water treatment. First of all, the experiments on the adsorption performance and biodegradability of the waste water are indispensable. Then, according to the result of static adsorption isotherms experiment on the raw water, the appropriate kind of the activated carbon can be chosen, and on the basis of dynamic adsorption isotherms experiment, the basic parameters can be determined. Ultimately, according to the process scale and condition of the field, BAC adsorption devices and its structure as well as supplementary equipment can be determined.

The activated carbon used in BAC process, should be highly developed in the pore structure, especially for the filter pores. Quality of the outflow is directly influenced by the filtering velocity, height of the carbon layer, the retention period and the gas-water ratio. In practice, the general filtering rate is 8-15 km/h. Retention period: According to the different pollutants, the general retention period should be 6-30 min; when the process is mainly removing smelly odor from the raw water, the period should be 8-10 min; when the process is mainly dealing with COD_{Mn}, the period should be 12-15 min. Gas-water ratio: As to aerobic microorganism, sufficient DO in the activated carbon layer is needed. Generally, DO>1 mg/L is proper in the outflow, therefore the design is based on a (4-6):1 gas-water ratio, specific details are based on the height of the carbon layer and concentration of organic contaminants. Generally, the thickness of the activated carbon layer is 1.5-3 m, which is determined by the leaping curve of the activated carbon. The growth of microorganisms and suspended solids brought by the inflow on the long-term operating biological carbon bed may block the carbon layer, and the biofilm on the surface of the activated carbon is unlikely to be discovered by naked eyes. Once the thickness is out of limits, the adsorption ability of the activated carbon will be affected undesirably. Therefore, the carbon layer should be washed periodically. Related parameters for backwash are shown in table 1.

Backwash type		Granular activated carbon grade	
		2.38~0.59 mm	1.68~0.42 mm
Air-water backwash	Water wash intensity [L/(m²·s)]	11.1	6.7
	Water wash interval (min)	8~10	15~20
	Air wash intensity [L/(m²·s)]	13.9	13.9
	Air wash interval (min)	5	5
Water wash and surface wash	Water wash intensity [L/(m²·s)]	11.1	6.7
	Water wash interval (min)	8~10	15~20
	Air wash intensity [L/(m²·s)]	1.67	1.67
	Air wash interval (min)	5	5

Table 1. Backwash parameters for activated carbon

2.2. O₃-BAC process and the evaluation of ozonation

2.2.1. Mechanism and characteristics of O₃-BAC process

In practice, there are still some problems when BAC technology is used alone, for example, some difficult biodegradable materials can not be removed effectively and the working life of BAC would be reduced. Meanwhile, in order to ensure the safety of water distribution system, disinfection is indispensable after biological treatment. When chlorine treating potable water is used, large amounts of halogenated organic byproducts will be produced during the reaction between the organics and the chlorine. Among those byproducts, THMs, HAAs, etc., are carcinogenic. Therefore, before BAC treatment, pre-ozonation, the O₃-BAC Process, is widely used, which concludes 3 procedures: ozonation, adsorption effect of activated carbon and biodegradation[19]. When O₃-BAC Process is used, organic will be firstly oxidized into small degradable molecules by strong oxidation of ozone, then the small degradable molecules will be adsorbed onto the activated carbon and degraded by microorganism, simultaneously the oxygen discomposed from ozone will enhance the level of DO, which makes DO in raw water be saturated or approximately saturated, which in turn, provides necessary condition for biodegradation[20-25]. Fig. 4. shows a simplified model of mutual effects among the main factors during O₃-BAC Process [26].

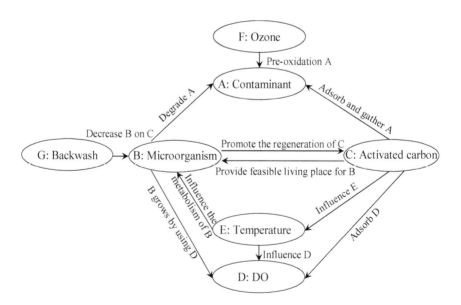

Figure 4. Model scheme of O₃-BAC

2.2.2. Effect of ozonation on molecule weight distribution and the molecule structure of organic matters[27-30]

2.2.2.1. Effect of ozonation on molecule weight distribution of organic matters

Pre-ozonation can change the properties and structures of organics in raw water. By using high performance liquid chromatography (HPLC), the variations of molecule weight before and after ozonation is studied, the result is shown in Fig. 5. As shown, the molecule weight mostly distributed within a range of 2000-6000, and 2000-3000 after the ozonation, the amount of organics with a relative molecule weight under 500 is increased dramatically. This indicates that macromolecules are in a high proportion in raw water, but after the ozonation, the proportion of macromolecular organic matter decreases while small molecule weight organics increases, which implies that part of the intermediate material from ozonation is namely the increased small molecular weight organic matter.

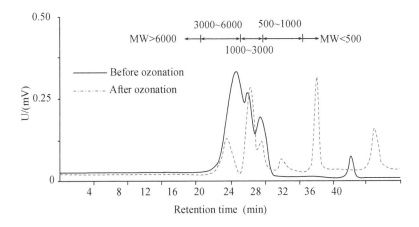

Figure 5. Molecular weight distribution before and after ozonation

2.2.2.2. Effect of ozonation on the structure of organic matters

In this part, Gas Chromatography-Mass Spectrometer (GC-MS) is used to analyze the structure of organics in raw water before and after ozonation, results are shown in Fig. 6. As shown, the organics in raw water are mainly aromatic hydrocarbons, chain hydrocarbon and aliphatic organics; after ozonation, the amount of aliphatic organic matters increases significantly and the amount of esters tend to increase too, which indicate that part of the aromatic organics are oxidized into organics of oxygen-containing groups, such as fatty acids, carboxylic acids and esters.

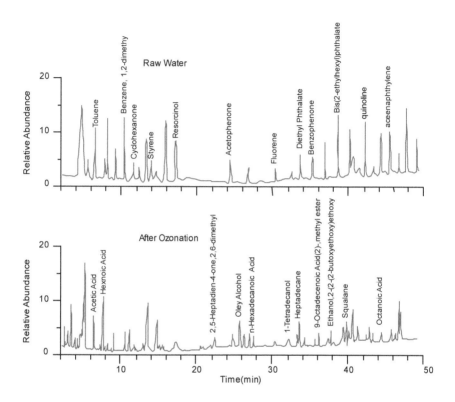

Figure 6. GC-MS chromatogram of raw water before and after ozonation

2.2.3. Improvement of biochemical properties of organics by ozonation

During the process of ozonation, complex chemical reactions occurred between ozone and organics. Pre-ozonation can change the biodegradability of organics in water, so generally BDOC is used as an index to analyze the water after ozonation. Fig. 7. shows the variation tendency of DOC, BDOC and BDOC/DOC in raw water after ozonation. The result indicates that after ozonation, DOC in raw water is seldom removed while BDOC increases obviously, among which the ratio of BDOC/DOC increases from 11.6 % to 26.4 %. After ozonation, the biodegradability of organics in water is highly improved, which enhance the biodegradation of BAC filter bed in the following procedure.

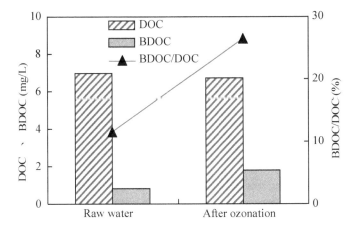

Figure 7. Variations of DOC, BDOC, BDOC/DOC before and after ozonation

2.2.4. Improvement of ozonation on biodegradability of organic matters

Relevant researches indicate that ozonation can change the molecule weight distribution, structures and the biodegradability of the organics in water. To thoroughly understand the role organics degradation and ozonation played in the whole process, degradation of the organics without dosage of ozone shall be taken for comparison. Variation of DOC, BDOC, BDOC/DOC on each layer of BAC bed with ozone or without ozone are both reviewed, the result is shown in Fig. 8 and 9.

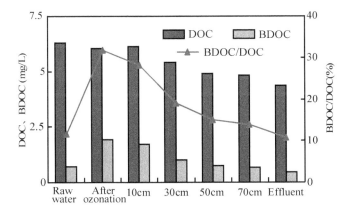

Figure 8. Degradation effects of different layers in BAC bed with ozone pretreatment

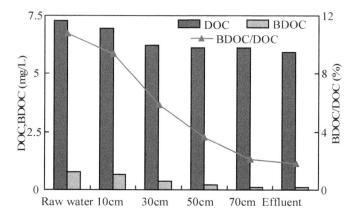

Figure 9. Degradation effects of different layer in BAC bed without ozone pretreatment

As shown in the figure, no matter the ozone is added or not, the tendencies of organics degradation in BAC bed are similar. With the addition of ozone, the properties of organics after ozonation will change, and the amount of BDOC increased. Viewed from aspect of general efficiency of removal, with the addition of ozone, the removal rate of DOC is 31 %, while 18 % removal rate without ozone.

According to further analysis of the data in Fig. 9, results show that, degradation efficiency of DOC without ozone is 18.4 %, DOC of outflow from the BAC bed is 5.92 mg/L and the removal quantity of DOC is 1.351 mg/L, in which the removal quantity of BDOC is 0.676 mg/L. In this case, the removal of NBDOC takes a proportion of 49 % in the total removal of DOC. Bio-degradation mainly removes BDOC in organics, while adsorption mainly removes NDOC and part of BDOC. That is to say, for BAC process absorption takes a proportion of 49 % during the entire removal process without dosage of ozone. Similarly, to analyze the data shown in Fig. 8, the variation of adsorption and bio-degradation effects during the organic degradation process whether added with ozone or not are shown in Table 2.

BAC/DOC Degradation	With Ozone Pretreatment	Without Ozone Pretreatment
Adsorption	35 %	>49 %
Biodegradation	65 %	<51 %

Table 2. Effect of ozonation on adsorption and biodegradation during BAC process

According to Fig. 3, on the aspect of the adsorption and bio-degradation in DOC removal process, ozonation can enhance the bio-degradation effect greatly, while exert a negative effect on adsorption of activated carbon.

3. Characteristics of microorganisms and mechanisms of pollutants degradation in BAC filtration

3.1. Immobilization of microorganisms on the carriers

At present, most of the BAC process is based on ordinary BAC which is naturally formed during the long-term operation. Due to the complexity of biofacies on its surface, the greatest deficiency of conventional BAC is that dominant microflora are hard to be formed, which has been improved by the rapid development of modern biology technology and the technology known as immobilization biological activated carbon (IBAC). The basic principle of IBAC is screening and acclimation dominant biocommunity from nature, followed by immobilizing the community on activated carbon, to enhance the efficiency and rate of degradation. Meanwhile, dominant biocommunity should be nonpathogenic and has strong antioxidant capacity and enzyme activity which enable the growing and reproducing under poor nutrition environment. Therefore, the effect of biodegradation after activated carbon adsorption saturation is highly improved by IBAC[31-32].

3.1.1. Intensification and immobilization of dominant biocommunity

Dominant microflora is selected by artificial screening and domesticating or directive breeding through biological engineering technology. Separation is the first step for getting dominant biocommunity. However, the filtered biocommunity may contain pathogenic bacteria, or with low-rated growth or high requirements for nutrition, which make it inadaptable for practical application, therefore, preliminary screening is indispensable for guaranteeing the security of selected bacterial strains. Due to the high complexity of biofacies, various species but less nutrition matrix in water, low-activated bacterial strain screened. This situation makes a relatively high disparity from the effective biological treatment. Therefore, intensification on selected bacterial strain is necessary. The process of intensification is actually the process of induction and variation of biocommunity. By changing nutritional conditions repeatedly, the bacterial strain will gradually adapt to the poor nutrition environment in fluctuation, so that the bacterial strain immobilized on activated carbon can preserve a strong ability for biodegradation[33].

During IBAC treatment, the process of microorganisms immobilizing is of great complexity, which is not only related to various acting forces, but also involves microorganism growth as well as the ability for producing extracellular and external appendages[34]. DLVO theory can give a better explanation that how bacteria are attached, when bacteria are regarded as colloidal particles[35-37]. Although there is a relatively high repulsive force during the process when the bacteria approaching to the activated carbon, the bacteria will ultimately contact with the activated carbon under the bridging effect of EPSs[38]. The bacteria may take advantage of the bridging effect with surficial particles, which enables the attachment of itself with filter materials at secondary extreme point in accordance with the theoretical level diagram of DLVO theory, rather than by the certain distance or overcoming the necessary energy peak as that of non-biological particle[39], see Fig. 10. for details. Although there are various ways to immobilize microorganisms, and any method with a limitation for free-flowing of microorganisms can be used to produce and immobilize microorganisms, an ideal and universal application way for immobilizing microorganisms is still not available. There are 4 common methods of immobilization, see Table 3 [40-42] for details.

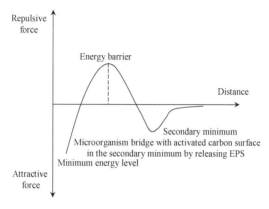

Figure 10. Total potential energy for express microbial immobility on activated carbon based on DLVO theory and short-range force

Property	Covalention	Adsorption	Covalent	Embedding
Implementing difficulty	Moderate	Easy	Difficult	Moderate
Binding force	Strong	Weak	Strong	Moderate
Active surface	Low	High	Low	Moderate
Immobilized cost	Moderate	Low	High	Low
Viability	No	Yes	No	Yes
Applicability	Bad	Moderate	Bad	Good
Stability	High	Low	High	High
Carrier regeneration	Unable	Able	Unable	Unable
Steric hindrance	Larger	Small	Larger	Large

Table 3. Comparison of the methods for immobilization of common microorganisms

3.1.2. Growth characteristics of dominant microflora on the surface of activated carbon[43]

In the early period of IBAC operation, the results for the biomass and biological activity on the surface of activated carbon monitored continuously are shown in Fig. 11. and Fig. 12.. At initiating stage, there is a rapidly dropping period for the dominant micoflora biomass on activated carbon. As time goes by, dominant biocommunity becomes steady gradually. At initial stage of BAC process, the biomass on activated carbon surface is in little, different from IBAC process, therefore, the microbial action can be ignored basically. During this period, variation of the biological activity for dominant biocommunity is similar to the change of biomass.

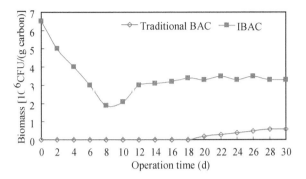

Figure 11. Variations of biomass on activated carbon

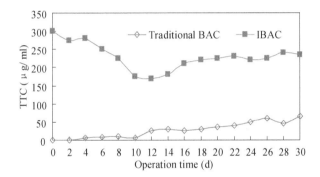

Figure 12. Variations of biological activity in two activated carbon processes

The variations of biomass and biological activity of microbe on activated carbon in long-term operation process are shown in Fig. 13. As shown, the biomass on carbon bed remains constant which mainly because of a gradual adaptation process of dominant biocommunity to the water, during which the adhesive ability of the dominant biocommunity is relatively low. Therefore microbe on the upper layer will be washed into the lower layer of the activated carbon, which has a beneficial effect on the reasonable distribution of the dominant biocommunity. With the extension of operating time, the dominant biocommunity will be adapted to the environment gradually. Meanwhile, due to the higher concentration of nutrient media in the upper layer, the biomass on the upper layer will be greatly increased and remains constant for a long period. Being different from the constancy of biomass on activated carbon, the biological activities in the upper and the lower layers of IBAC bed have a decreasing tendency as operation time goes. Shown from the result of PCR-DGGE, this reduction is mainly caused by the continuous incursion of the native bacteria with low capacity.

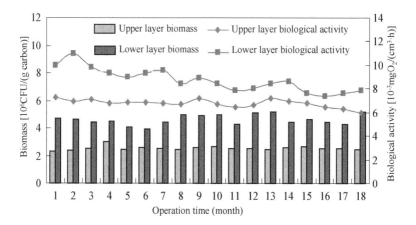

Figure 13. Variation of biomass and biological activity on activated carbon

3.1.3. Influencing factors of dominant microflora biological stability

During the IBAC process, the biological stability of the dominant microflora is the key factor to ensure an effective operation. However, the factors including the property of the activated carbon, the dosage of ozone, hydraulic retention time and the condition of backwash all have impacts on the dominant biocommunity at different levels. When all combined influences considered, the main factors of the activated carbon effects on the stability of dominant biocommunity are the distribution of pores in activated carbon, the physical and mechanical properties as well as its chemical property, among which molasses value is the primary controlling index [44-46]. Although ozonation may improve the biodegradability of water, provide an oxygen-rich condition and reduce the incursion of undesirable bacteria, as well as enhance the biological activity of dominant biocommunity, however, relevant research shows that when the residual ozone reaches even over 0.1 mg/L in the inflow of IBAC system, a restrain on dominant biocommunity will be shown[47]. By increasing the contact time of the dominant microflora with organics, the adsorption and mass transfer of organics can be enhanced, thus the biological activity of the dominant biocommunity is improved. Generally, the most appropriate retention time is about 20 min[48]. The backwash step has an impact on the biological activity of the dominant biocommunity, which makes biological activity of the community in each layer rapidly decreases from the normal level to the minimum, but still remains in a certain range. After the sharp cutoff, the biological activity of the community will recover at normal level rapidly. Meanwhile, the intensity of backwash and its time also have a great impact on biological stability[49].

3.2. Characteristics of the biocommunity structure and distribution

The huge superficial area and rich porosity of the activated carbon provide a preferable habitat for microbe and prompt the formation of biological membrane. According to the research, the synergy effects of activated carbon adsorption and biodegradation enable O₃-BAC to realize the removal of organics in water (see 3.4 for details). Meanwhile, as the duration of BAC use gets longer, the action of organisms plays a more and more predominant role[50,51]. Quite a few researchers point out that the range of bacteria using the activated carbon as a carrier to habitat includes aerobic, anaerobic and facultative bacteria[52,53]. In this chapter the variation of bacteria communities and the characteristic of the community structure will be discussed, and the system chosen here is down-flow O₃-BAC system shown in Fig. 14.

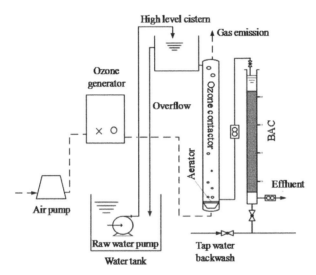

Figure 14. Pilot test setup

3.2.1. Characteristics of strain distribution

Various kinds of strains exist in BAC column (see Fig. 15). The picture demonstrates, (L-R), 4 sections (1, 2, 3 and 4) from top to bottom of the down-flow BAC layers respectively. The shadows in the finger prints represent different strains, and the numbers of the shadow zones represents the numbers of strains, while every shade represents a kind of strain, so the different shades represent the number of species indicatively. Fig. 15. shows that the number of shadow zones is 16 that can be discovered obviously, which indicates that the number of strains exist in BAC column is 16 according to detection. Meanwhile, along the carbon layer from top to bottom of the picture (L-R), there are 7 shadow zone shades turning weak gradually, which indicates that with the increasing of the carbon layer depth, these strain kinds are reduced gradually. There are 5 shadow zone shades turning darker, which indicates that with the increasing of carbon layer depth, these strain kinds increase. Another 4 zones remained do not exist in every section, among which the quantity of one kind of strain is not changing, 2 kinds of strains only exist in the upper side of the activated carbon, while only 1 strain exists in the bottom of the activated carbon bed.

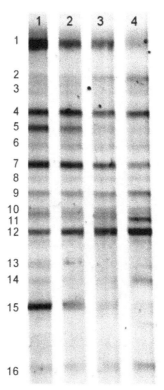

Figure 15. Analysis results of PCR-DGGE

3.2.2. Characteristics of biomass distribution

The growth of microorganism in BAC bed can be divided into three phases: logistic phase, stable phase and decayed phase (see Fig. 16). In the first period, microbe multiplies. In the initial period of biofilm formation, microbe quantity is rather low, and the highest biomass only reaches 70nmolP/gC. When the water temperature is proper and the inflow contains a lot organics, biological multiples rapidly. After one and half a month of the reduction period, it reaches the stable phase. Generally, in this phase, the concentration of the substrate declines, which means the degradation rate is rather high; DO consumption is also in large amount. In the second phase, microbe grows stably. After a period of cultivating when the water temperature is proper, the microbe basically reaches a state of stability, in which the microbe quantity maintains in the range of 450-520nmolP/gC and the microbe grows normally. Meanwhile, the removal effect on pollutants in water is also steady, and at this point, the eco-system of BAC bed is relatively steady. This phase is characterized by equilibrium between biological membranes of new cells and the loss of membranes causing by physical force. In the decay phase, when water temperature as well as other conditions varies fiercely, the microbe quantity decreased rapidly and entering the final phase, in which degradable rate decreased, quality of outflow will be worsen.

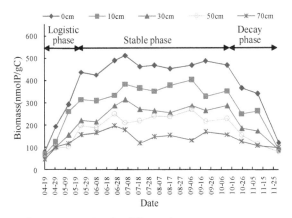

Figure 16. Variation of biomass in BAC Bed at different time

As the substrate is oxidized and degraded by the microbe membrane from top to bottom, the concentration decreased gradually, lead to the decline of available organic concentration and the poor nutrition condition of the microbe. Besides, it's also related to DO concentration. DO distribution in BAC column is gradually decreased from top to bottom, the lower it goes on the carbon bed, the fewer the available DO for microbe to reproduce, so does the biomass. Thus, the biomass distribution in BAC bed shows a feature of gradually decreasing from top to bottom (see Fig. 17).

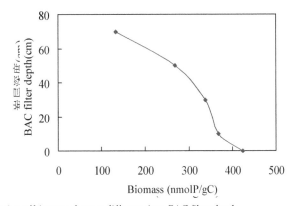

Figure 17. Variation of biomass along at different time BAC filter depth

3.2.3. Characteristics of biocommunity distribution

For down-flow BAC bed, though there was a decreasing tendency of bacteria adsorbed on activated carbon from top of the carbon layer to the bottom , the variation tendencies of ammonifying bacteria, anti-vulcanization bacteria and denitrifying bacteria are not exactly the same, among which ammonifying bacteria showed a decreasing tendency from top of the carbon layer to the bottom; anti-vulcanization bacteria showed an increasing tendency from top of the carbon layer to the bottom reaching a steady quantity; while denitrifying bacteria which located higher than 30cm of the carbon layer showed a gradual increasing tendency, however, after reaching the critical value of 30cm, it would decrease rapidly (shown in Fig. 18). As whole, aerobic bacteria is dominant in BAC bed.

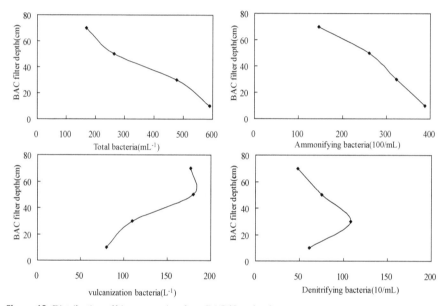

Figure 18. Distribution of biocommunity along BAC filter depth

As ammonifying bacteria is strictly aerobic, and denitrifying bacteria is amphitrophic, while anti-vulcanization bacteria is strictly anaerobic, the result shown in Fig. 18 is conducted by effects of two sides. Firstly, BAC is a kind of biofilm in essence, the thicker it gets, the more likely to create an anaerobic or hypoxia environment internal, which directly induces the existence of strictly anaerobic bacteria and amphitrophic bacteria. Secondly, variation of DO in activated carbon bed changes the distribution of biocommunity. Because aerobic bacteria is dominant in BAC bed, in the down-flow system, there is a decreasing tendency along the carbon layer (shown in Fig. 19), although, the concentration of DO in bottom of the layer is about 2-3mg/L, the flow inside BAC bed is laminar flow, which is not easy for the transmission of oxygen, hence, the bottom of activated bed shows anaerobic or anoxic characteristic, resulting in existence of amphitrophic bacteria and anaerobic bacteria.

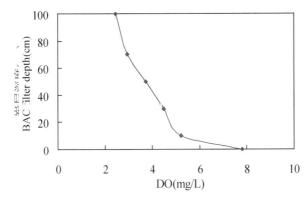

Figure 19. Variation of DO along BAC filter depth

During a steady operation of BAC process, a complex ecological system consists of bacteria, fungi, and algae which is from protozoa to metazoa is formed, which indicates the biofacies in BAC bed is of great abundance, and the ecological system of great diversity is fully developed which has a strong ability of avoiding the load impact. Part of the typical organisms inside BAC bed examined by microscope is shown in Fig. 20[54].

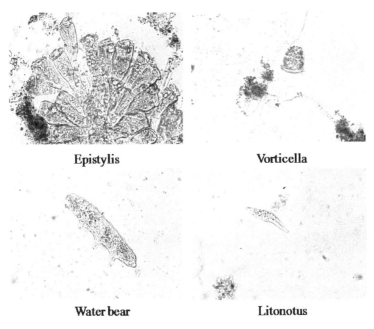

Figure 20. Microscope examination of microorganisms in BAC filter

3.2.4. Biocommunity diversity [55]

By collecting the biofilm on the surface of BAC, taking the total DNA of microbe and building a cloning library of 16S rDNA bacteria, as well as choosing a random one to clone with and conducting a DNA sequencing, the most similar bacteria to the cloned one can be determined after comparing with the community recorded in the database. Without relying on the domesticating method, the biocommunity's structure can be directly analyzed.

As shown in Table 4, in the biocommunity, α-Proteobacteria is dominant, β-Proteobacteria is the second category of this ecological system and δ-Proteobacteria is the third. Meanwhile, Planctomycetes bacteria also take a big proportion in this gene library.

Classification	Proportion (%)
α-Proteobacteria	26.5
β-Proteobacteria	16.3
γ-Proteobacteria	2.0
δ-Proteobacteria	16.3
Nitrospira	2.0
Planctomycetes	12.2
Bacteroidetes	2.0
Gemmatimonadetes	6.1
Acidobacteria	4.1
Unclassified	—
Proteobacteria	—
Unclassified Bacteria	10.2
Actinobacteria	2.0

Table 4. Fraction of different bacteria in gene clone library

After entering the sequence of genotype into NCBI website and comparing it by BLAST procedure with the existing sequence, it can be concluded that many of the bacteria's 16S rDNA sequencing has a rarely similarity with the existing ones in database, among which many have a similarity below 95%, reaching 88% the minimum, and most of the sequencings are from environment like soil, activated sludge, underground water, rivers, lakes and the urban water supply system.

Proteobacteria and other phylogenetic trees (shown in Fig. 21 and Fig. 22) were built to further understand the status of bacterial system development and assure the species of the cloned bacteria. As shown in Fig. 21 and 22 in samples, most of the cloned ones are similar to the bacteria which are not cultivated, only cloned 1-22 shares the same species with Chitinimonas taiwanensis in the phulogenetic tree.

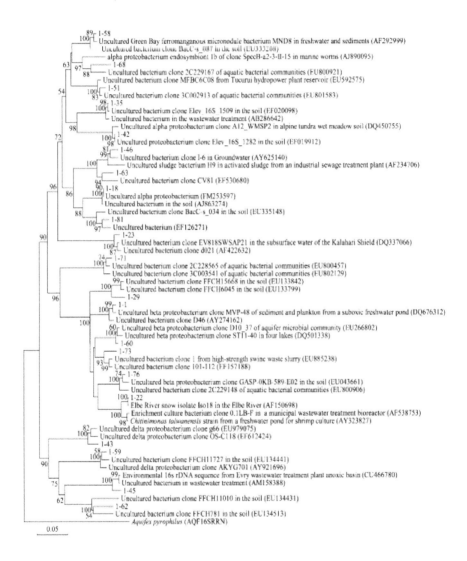

Figure 21. The Phulogenetic tree of Proteobacteria

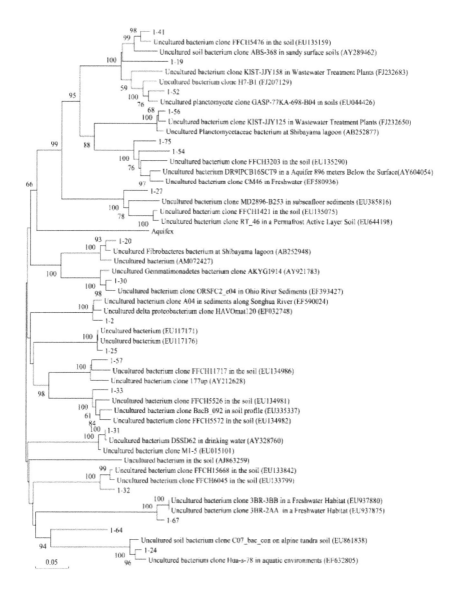

Figure 22. Phulogenetic trees of other bacterial system in carbon samples

3.3. Synergy of activated carbon adsorption and biodegradation

3.3.1. System construction and research method

The adopted system device is shown in Fig. 23., ozone is dosed via titanium microporous aerator from the bottom column of contact reaction to realize the contact of gas and water in a reversal way. Generally the dosage of ozone is 1.5mg/mgTOC, and the retention time is 11 min. After ozonation, the outflow will equivalently enter the bio-ceramsite filter column (the diameter of ceramic particles is 2-3mm) and bio-activated carbon filter column through the high level cistern and the control valve respectively. Two filter columns are both made of organic glass with 70mm inner diameter and 1650mm height, and the effective height of each filer is 1050mm. For each column, there are sample outlets at height of 100mm, 300mm, 500mm and 700mm along the top of each filter. The empty bed contact time for each column is 15min, and the cycle period of backwash is 3d. Under the same condition, by analyzing the removal effect of BAC and BCF towards DOC and BDOC to conduct quantitative analysis of organics removed by BAC adsorption and biodegradation, the theory on removal of pollutants in water by BAC will be discussed and the removal induced by the synergy effect of BAC adsorption and biodegradation[56,57] will be determined accordingly in this part.

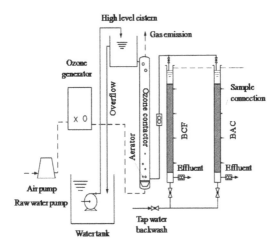

Figure 23. Pilot test arrangement

3.3.2. Quantitative calculation method of relationship between adsorption and biodegradation in BAC filter bed

3.3.2.1. Quantitative calculation method

This part compares O_3-GAC with O_3-BCF in which the biodegradation is dominant. By determining the iodide adsorption by BCF, only to find out the iodide adsorption is nearly zero, which may conclude that for BCF filter bed the removal of organics in water is mainly realized by biodegradation, and the BCF adsorption is extremely faint. Assuming the quantity of biodegradation is in proportion to biological activity, the more active the microbe is, the more organics degraded will be. Therefore, determining the quantity of organics being degraded by BAC can be conducted by calculating the ratio of BAC biological activity with BCF biological activity, which can also determine the roles played by BAC filter bed adsorption and biodegradation in the process of organic removal. The calculation in details is shown in Fig. 24. Shown from the simplified calculation diagram in Fig. 24, two kinds of indexes need to be assessed to quantitatively determine the relationship between BAC adsorption and biodegradation within the process of organics removal:

1. to evaluate the comprehensive effect of BAC and BCF on organics removal, the system is characterized by DOC and BDOC;
2. to compare and analyze the biological activity of carriers from each filter bed.

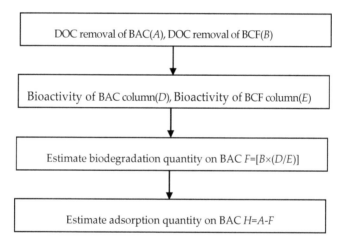

Figure 24. Illustration of quantification of the removal of organic matters by adsorption and biodegradation

3.3.2.2. Analysis of biological activity

The basic idea of in situ substrate uptake rate detection method[58,59] is to converse time-variant variables into space-variant variables. First assumption is to consider matrix concentration as a differential function of filter bed height; the second assumption is that under proper conditions, any of the micro-unit taken from the inner-filter bed can achieve stability. Through the total differential method or the law of conservation of mass, the final expression of matrix degradation velocity can be determined, at this point, the law of conservation of mass is applied, for its intuitionism.

As shown in Fig. 25, under a constant living environment for microbe, in situ oxygen uptake rate (ISOUR) can be described as the oxygen consumption of microbe grown on filter materials per unit time and volume (written as R, mg/(L·h)). Consider the filter porosity in filter column as e, mass concentration of DO in inflow and outflow are $\rho(DO_i)$ and $\rho(DO_e)$ respectively, taking ωdh as the micro-unit taken from depth of h in bio-filter bed (ω is section area of filer column, m²), and the variation in DO caused by the effect of microbe inside the micro-unit can be described as $R(1-e)\omega dh$, according to the law of conservation of mass:

$$(\rho(DO_i) - \rho(DO_e))Q - R(1-e)\omega dh = 0 \tag{1}$$

ISOUR can be described as

$$\frac{dC}{dt} \tag{2}$$

In this equation: Q stands for the quantity of water through the filter column, m³/h; v stands for the filter velocity, m/h;

$$\frac{d\rho(DO)}{dh}$$

stands for the gradient of DO at a height of h, along the filter variation height curve. As far as the aerobic biological treatment concerned, the biodegradation velocity on organic matrix is in proportion to its ISOUR. With abundant DO in BAC and BCF filter columns, the breeding microbe is mostly aerobic bacteria. Therefore, by calculating the oxygen consumption rate in the process of treating organics by BAC and BCF filter columns, the respective biological activity can be determined to analyze and evaluate the effect of biodegradation of BAC column.

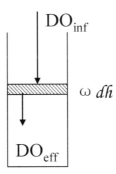

Figure 25. Calculation model of ISOUR

3.3.3. Quantitative comparison between relationship of adsorption and degradation in BAC filter bed

3.3.3.1. Analysis of biological activity (ISOUR)

Variation of DO in each section of BAC and BCF is shown in Fig. 26, according to equation (3-2), the biological activity of each section can be determined, and the result is shown in Figure 27. As can be seen from Fig. 26 and 27, an obvious change in biological activity along with the various depth of carbon layer exists in both BAC filter bed and BCF filter bed, showing a gradual decreasing tendency along with the depth, which is closely related to the distribution of biomass on carriers. Biological activity ratio of BAC to BCF can be determined based on Fig. 27, shown as D/E in Fig. 24, and the result is shown in Fig. 28.

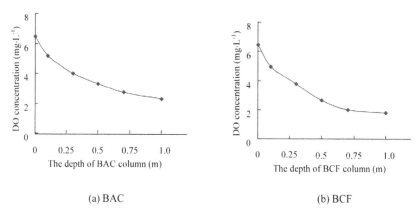

(a) BAC   (b) BCF

Figure 26. Variation of DO along BAC and BCF filter

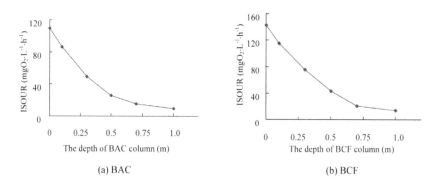

(a) BAC   (b) BCF

Figure 27. Variation of microbial activity along BAC and BCF filter

Fig. 28 indicates that the microbial activity of BCF filter bed is higher than that of BAC bed, and D/E for each section is ranging from 0.60 to 0.77. Among those sections, four of them have a ratio higher than 0.7, hence D/E=0.72 is set in quantitative calculation.

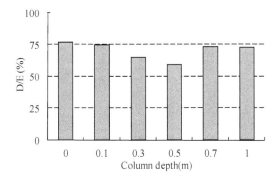

Figure 28. Ratio of microbial activity in BAC and BCF section (D/E)

3.3.3.2. Analysis of the organic removal effect of BAC and BCF

The respective removal effect of BAC filter bed and BCF filter bed on DOC and BDOC after ozonation is shown respectively in Fig. 29 and Fig. 30. Only a few DOC, which accounts for 4% of raw water is removed by ozonation. For calculation, the removal of DOC and BDOC shown in Fig. 29 and Fig. 30 is in relative terms with ozonation. Analysis of the statistics of Fig. 29 and Fig. 30 is shown in Table 5, in which NBDOC stands for non-biodegradable dissolved organic carbon. As shown in Table 5, the removal effect of BAC on DOC is mainly focused on removing BDOC, among which removal of BDOC accounts for 85% of the degradation DOC, while for BCF filter bed, this proportion can be as high as 98%.

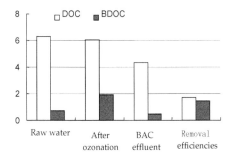

Figure 29. Results of DOC and BDOC removal in BAC filter

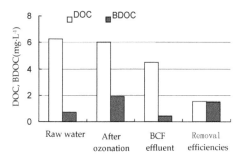

Figure 30. Results of DOC and BDOC removal in BCF filter

Column	DOC removal	BDOC removal		BDOC/DOC (%)
		NBDOC	BDOC	
BAC	1.685	0.234	1.451	85
BCF	1.525	0.03	1.495	98

Table 5. Calculation results of DOC and BDOC removal in BAC and BCF filter

3.3.3.3. Quantitative analysis of O₃-BAC adsorption and biodegradation

According to the calculation method in Fig. 24, and considering the analysis of BAC, BCF biological activity and their effects on removing organics, Table 6 shows quantitative calculation results of organic matter removal by adsorption and biodegradation in BAC filter. To intuitively understand how O₃-BAC biodegradation and adsorption work, Fig. 31 along with the analysis of Table 6 and Fig. 29, shows a simplified model of O₃-BAC removes organics.

Column	DOC removal	BDOC removal	
		NBDOC	BDOC
Adsorption	0.595	0.234	0.361
Biodegradation	1.09	0	1.09
Total	1.685	0.234	1.451

Table 6. Calculation results of organic matter removal by adsorption and biodegradation in BAC filter

As shown in Fig. 31, it is conducted by the synergy effect of activated carbon adsorption and biodegradation of O₃-BAC. Under the system research conditions, BAC bed biodegradation is dominant which accounts for 65% removal of total organics, furthermore, this 65% is mainly organics of BDOC, which indicates that the main organics removed by biodegradation is readily degradable dissolved organics, which can be concluded that biodegradation has a remarkable selectivity. In contrast, activated carbon adsorption plays a supporting role in the system, while via adsorption the quantity of removed difficult biodegradable material is roughly similar to the quantity of the readily one[60].

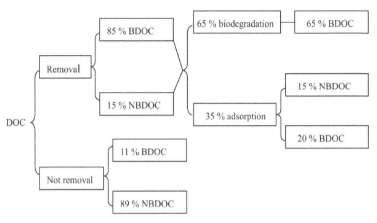

Figure 31. Illustration of organic matter removal by BAC filter after ozonation

3.4. Degradation kinetics of pollutants in BAC bed

3.4.1. Kinetics model establishment

Adopted system device is shown in Fig. 32. The relationship among microbial growth, the initial concentration of microbe and that of the substrate is the key factor to establish the degradation kinetics model. The relationship can be reflected by various models and the Monod equation is universally acknowledged as the practical model[61]. For water treatment field, the specific degradation velocity of the substrate is more practical than the specific growth velocity of microbe. Considering the specific degradation velocity of the substrate according to the physical law, the equation below is founded[62]:

$$-\frac{dC}{dt} = v_{max} \frac{XC}{K_s + C} \qquad (3)$$

Wherein, $\frac{dC}{dt}$ stands for the degradation velocity of organic substrate; C stands for residual organic substrate concentration in mixed liquor after a reaction time t; K_s stands for saturated constant; v_{max} stands for the max specific degradation velocity of organic substrate.

In the equation, C is the key factor for calculating kinetic equation. When $K_s > C$, the equation below is founded:

$$-\frac{dc}{dt} = k_1 XC \qquad (4)$$

When $K_s < C$, the equation below is founded:

$$-\frac{dc}{dt} = k_2 X \qquad (5)$$

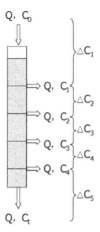

Figure 32. Illustration of material balance

Taken X (biomass) as horizontal coordinate and $\frac{dC}{dt}$ as vertical coordinate, if the diagram obtained is linear, the model belongs to high matrix organic degradation, while if the diagram is in exponential form, the model belongs to low matrix organic degradation.

As both BAC and BCF have a homologous microbial characteristic, by drawing an analogy with the degradation model of BCF, the BAC biodegradation effect can be described. According to the experimental system, material balance can be established which is shown in Fig. 32.

As shown in Fig.32, according to material balance law, along the section, the equation below is founded:

$$QC_0 = Q\left(C_1 + \sum_{i=1}^{1}\Delta C_i\right) = Q\left(C_2 + \sum_{i=1}^{2}\Delta C_i\right) = Q\left(C_3 + \sum_{i=1}^{3}\Delta C_i\right) = Q\left(C_4 + \sum_{i=1}^{4}\Delta C_i\right) \tag{6}$$

In which ΔC can be calculated by each C of all the sections, Δt can be calculated by the height of filter through each section and the filter velocity; the biomass of each BCF section represented by X can be measured, hence, taking biomass as the horizontal coordinate, $\Delta C_i/\Delta t$ of each section as the vertical coordinate to conduct linear regression analysis so as to determine the type of biodegradation model.

3.4.2. Characteristic of BAC degradation towards organics

3.4.2.1. Characteristics of BCF bed degradation

Fig. 33 shows the monitoring result of biomass in different BCF bed sections during a 7-month operation period. It can be seen from Fig. 33 that less biomass initially, till 20th May, the biomass of each section increased obviously and being steady gradually, which symbolically means it is mature for biofilm colonization in bio-ceramsite filter bed. However, from the middle of October, biomass of each section is decreasing rapidly, which is partially induced by the gradual reduction for water temperature during operational period. Therefore, it is appropriate for choosing the data recorded from 20th May to the middle of October to be compared with. Fig. 34 shows the BDOC variation of each BCF bed section in this period. Fig. 34 suggests that organic concentration of each section is relatively steady and decreases along the direction of water flow, as well as gradual reduction between the differences of organic concentration from each section.

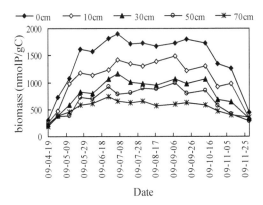

Figure 33. Variation of biomass along BCF bed

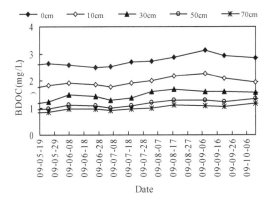

Figure 34. Variation of BDOC along BCF bed

By choosing the biomass of each section in the duration from 20th May to the middle of October and the mean value of BDOC variations to analyze the relationship between the biomass and $\Delta Ci/\Delta t$ variation according to the built model, the result is shown in Fig. 35. From Fig. 35, BCF bed biodegradation is in compliance with high matrix featured degradation model.

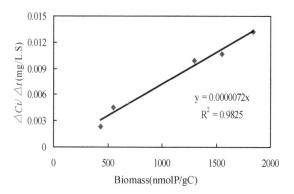

Figure 35. Calculation of BDOC degradation model in BCF column

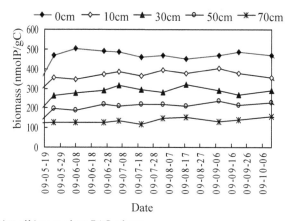

Figure 36. Variation of biomass along BAC column

3.4.2.2. Characteristic of BAC bed degradation

The variation of biomass in BAC bed at the same period is shown in Fig. 36. Fig. 36 shows that the biomass of BCF is much more than that of BAC, considering from the calculation above, we can find that for BCF system, biodegradation is in compliance with high matrix kind. Hence, it is the same with BAC system. Considering the two matrixes share the same homologous microbial, we can establish the BAC biodegradation equation by employing the calculated K:

$$\Delta C_i / \Delta t = 0.0000072 X_i \tag{7}$$

Wherein ΔC_i stands for the difference between the organic substrate concentration in mixed liquor residual of section i and that of section i-1; Δt stands for hydraulic retention time from section i to section i-1; X_i stands for biomass of section i.

3.4.2.3. Analysis of BAC bed organics degradation

Fig. 37 shows the analysis of the monitored variation of organics in each section of BAC bed at the same period, from which, organics concentration along the direction of water flow is decreasing gradually. However, as discussed previously, variation of organics is not only conducted by biodegradation, among which activated carbon adsorption is of the same importance. According to the biodegradation model mentioned previously, via calculation, BDOC removal conducted by microorganism biodegradation of each section as well as organics quantity adsorbed by activated carbon can be defined as $\triangle$BDOC.

It can be seen from Fig. 37 that as the carbon layer goes deeper, the biomass become less gradually, also a decreasing in biodegradation and the BDOC quantity which is degraded by microbe, while at this point, the adsorption increases gradually.

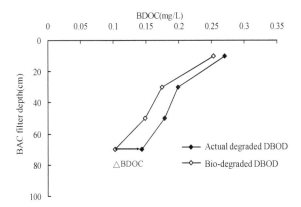

Figure 37. Calculation results of organic matter removal by degradation in BAC bed

4. Study on the safety of water quality by O_3-BAC process

4.1. Safety of microorganism

Because of ecclasis of microorganism and hydraulic erosion, microorganism may release from activated carbon bed, which will cause a series of water quality safety problems. Therefore, the microorganism safety of O_3-BAC process is very important for its application.

Microorganism safety of O_3-BAC technology should include the following four aspects[63], which characterize microorganism safety from different aspects respectively. Firstly, pathogenic microorganisms and toxic substances produced by their metabolism are the core problems concerning biological safety. Secondly, biocommunity mainly including bacteria, protozoa and metazoan is the expansion of the first aspect, which is based on some kinds of correlativity relationships between the biocommunity and pathogenic microorganisms. Thirdly, there is a certain connection between water quality parameters such as turbidity, biological particle number and the risk of pathogenic microorganisms. Fourthly, Assimilable Organic Carbon (AOC) is one of the indexes characterizing bacteria regrowth potential. Low AOC indicates a small possibility for bacteria regrowth and low pathogenic microorganisms risk[64].

Correlations exist among those four aspects. Among them, the first and second aspects are relatively intuitive, and have intimate connections with microorganism safety, but not easy to be detected. The third and the fourth aspects are indirect indexes that can be detected quickly and easy to be automatically controlled, which is particularly important to the operation management of water plant.

According to the research and operation practice, O₃-BAC process produced abundant biocommunity, but pathogenic microorganisms were not found in the activated carbon and so were significant pathogenic microorganisms in effluent as well[64]. The effluent turbidity maintained under 0.1NTU on the whole, but it may beyond the standard during the preliminary stage (0.5~1h), the later period or the whole procedure of activated carbon bed filtration. Particle number was generally less than 50/ml in outflow, when the system worked stably, thus the microorganism safety can be guaranteed. However, it would reach up to 6000 per milliliter in primary filtration water. On the other hand, the effluent AOC basically maintained under 100 µg/L. According to relevant research result, AOC concentration in 50~100 µg/L could restrain the growth of colibacillus.

On the whole, microorganism safety problems have not been found in O₃-BAC process until now, while it must arise enough attention. This problem should be controlled by optimizing design parameters, strengthening the operation management and developing new treatment technology.

Ozonation contact reactor should be set up with more ozone dosing points to guarantee the removal effect of cryptosporidium and giardia insect, and CT value should be greater than 4 generally. The thickness of the activated carbon bed should be greater than or equal to 1.2m in general. In order to guarantee microorganism safety, a sand layer with certain thickness shall be considered to be added under the carbon layer. Besides, setting up reasonable operation period for biological activated bed and strengthening the management of initial filtrated water. If possible, water quality monitoring of every single biological activated carbon processing unit should be taken into account, for example, setting online monitoring equipments for turbidity and grain number. Furthermore, it can be effective by using other processing technologies or combining O₃-BAC with other technologies to solve this problem. For instance, reversed O₃-BAC and sand filter or O3-BAC and membrane filtration hybrid process (UF and MF) can be applied[65].

4.2. Safety of water quality

In addition to microorganism safety, toxic and hazardous compounds may be produced in actual operation of O₃-BAC process. Those toxic and hazardous compounds can be divided into two parts in general, namely ozonation byproducts and biodegradable byproducts respectively.

4.2.1. Ozonation byproducts

Ozonation byproducts can be divided into two kinds according to its origin, which are produced under the condition of humic substances and bromide existing in water. The former is caused by the reaction of O_3 and hydroxyl radicals, and the latter is caused by hypobromous acid. Composition of byproducts would become more complicated when ammonia nitrogen and amino acid exist at the same time[66].

4.2.1.1. Byproducts caused by humic substances

HA is the main composition of NOM in water, the molecular weight of material after ozonation is lower than that of NOM, and contains more oxygen during the reaction. The main byproducts are shown in Table 7. Among these byproducts, carbonyl compounds especially aldehydes should be paid most attention. Animal experiments indicated that formaldehyde, acetaldehyde, glyoxal and methylglyoxal have acute toxicity as well as and chronic toxicity, and furthermore, tube test also implied that these substances have genotoxicity, carcinogenicity, mutagenicity in different levels.

Carbonyl compound	Aldehydes	Aliphatics	Formaldehyde、Acetaldehyde、Propionaldehyde
		Aromatic	Benzaldehyde
	Dialdehyde	Aliphatics	Glyoxal 、 Methylglyoxal 、 Maleic aldehyde 、 Furaldehyde
	Ketones		Acetone
Oxygenous carboxylic acid			Aldehydoformic acid、 Methylglyoxal、 Diethyl Ketomalonate、
Carboxylic acid	Monocarboxylic acid		Formic acid、 Acetic acid ~$C_{29}H_{59}COOH$
	Dicarboxylic acid		Oxalic acid、 Maleic acid、 Galactaric acid、 Fumaric acid
	Acromatic carboxylic acid		Benzoic acid、 Phthalic acid
Bioxide			Hydroquinone、 Catechol
Other			Heptane、 Octane、 Toluene

Table 7. Main ozone byproducts owing to NOM

4.2.1.2. Byproducts caused by bromine

When bromine exists in water, bromine will be oxidized into hypobromous acid by ozone, and then hypobromous acid will be oxidized into bromate, which has carcinogenicity to human body[67]. When ammonia nitrogen and amino acid exist at the same time in water, the reaction speed of bromate and ammonia nitrogen is faster, thus, dosing little amounts of ammonia would often restrain the production of organobromine compounds. Moreover, when bromine concentration is higher in the water, cyanogen bromide becomes the main ozonation byproduct[68]. Byproducts caused by bromide are shown in Table 8.

	Bromoform	$CHBr_3$
By-products caused by bromides	Bromoacetic acid	Bromacetic acid Dibromoaceticacid Tribromoacetic acid
	Bromohydrin	Bromine sec-butanol
	Acetonyl bromide	Bromopropanone Dibromoacetone
	Hypobromous acid Hypobromite	HBrO、MeBrO
	Bromic acid、Bromate	$HBrO_3$、$MeBrO_3$
By-products under the condition of ammonia nitrogen coexist	Bromopicrin	CBr_3NO_2
	Dibromoacetonitril	$CHBr_2CN$
	Cyanogen bromide	BrCN
	Bromine amine	NH_2Br、$NHBr_2$、NBr_3

Table 8. Main ozone by-products owing to Br⁻

Ozonation byproducts could be removed by adsorption of activated carbon. Meanwhile, ozonation and hypermanganate hybrid method will reduce the amount of byproducts. Taking more ozone dosing points can shorten the mean contact time and reduce the residual ozone concentration, thus bromate production will be decreased accordingly. Moreover, dosing acid to lower pH, ammonia, excessive H_2O_2 and OH scavenger to water may reduce BrO_3^- production[69] as well.

4.2.2. Bio-degradation byproducts

The microorganisms in the BAC will produce some microbial products such as SMP (soluble microbial products), EPS (extracellular polymeric substances), etc., when microorganisms degrade organics in wastewater.

Most of such substances are difficult to be degraded and hazardous to human. When drinking water is treated by O_3-BAC advanced treatment process, the degradation byproducts in a certain concentration will be produced, which causes health risks by drinking water. The disinfection byproducts will be formed through difficult biodegradation byproducts during chlorination, which are complex, various and more harmful to human body. The microbial degradation byproducts are organic, thus these substances may lead to the growth of bacteria in the pipeline. Moreover, the microbial degradation products will result in membrane clogging, and reduce the membrane flux as well as the membrane life during the O_3-BAC and membrane filtration hybrid process to produce high-quality water[10].

Although the effluent of BAC may contain some microbial byproducts, the content is limited and is not harmful for human health. Therefore, controlling the ozonation byproducts is more important and meaningful.

Author details

Pengkang Jin, Xin Jin, Xianbao Wang, Yongning Feng and Xiaochang C. Wang
School of Environment & Municipal Engineering, Xi'an University of Architecture & Technology, Xi'an, P. R. China

5. References

[1] He Wenjie, Li Weiguang, Zhang Xiaojian, Huang Tingling, Han Hongda. Novel Technology for Drinking Water Safety [M]. China Architecture & Building Press. 2006

[2] Weissenhorn FJ. The Behavior of Ozone in the System and Its Transformation [J]. AMK-Berlin, 1977, (2): 51-57.

[3] Parkhurst, J. P. et al. Pomona Activated Carbon Pilot Plant [J]. J. WPCF, 1976, 37 (1).

[4] Li Weiguang, Ma Fang, Yang Xianji, Zhao Qingliang, Wang Qingguo. A Study on Purification Performance of Biological Activated Carbon [J]. Journal of Harbin University of Civil Engineering and Architecture, 1999, (32) 6: 105−109. (In Chinese)

[5] Eberhardt M S. Untersuchunger Zur Berwendung Biolosisch Arbeitender Ativkohle Gilter Bei Der Trinkwasseranf Bereitung[J].Wasser-Abwasser, 1975, 116(6): 29-34.

[6] Sontheimer H. The Mulheim Process. J. AWWA, 1978, 70(7): 62-68.

[7] Greening F. Experience with Ozone Treatment of Water in Switzerland. In: Andrews G F, eds. Proceedings 8th Ozone World Congress. Zurich: IOA, 1987, 49-54.

[8] Gomella C. Ozone Practices in France. J. AWWA, 1972, 64(1): 41-45.

[9] Tian Yu, Zeng Xiangrong, Zhou Ding. Development of Coordinating Technology for Ozonation and Biological Activated Carbon System [J]. Journal Of Harbin Institute Of Technology, 1998, 30 (2): 21-25. (In Chinese)

[10] Gao Naiyun, Yan Min, Le Yuesheng. Strengthening Treatment Technology for Drinking Water [M]. 2005.(In Chinese)

[11] Qing Tian, Jihua Chen. Application of bioactivated carbon (BAC) Process in Water and Wastewater Treatment [J]. Environmental Engineering. 2006, 24 (1): (84-86). (In Chinese)

[12] Shucheng Lan. Biological Activated Carbon Technology and Its Application for Wastewater Treatment [J]. Water & Waste Water Engineering. 2002, 28 (12): 125. (In Chinese)

[13] Xiaochang Wang. Theoretical and Technical Aspects of Ozonation in Drinking Water Treatment [J]. J. Xi'an Univ. of Arch. & Tech. 1998, 30 (4): 307-311. (In Chinese)

[14] Liang Xu, Wenmei Jiang, Yu Zhang, Jinguan Liu. Progress Research and Application of Biological Activated Carbon Technology in Water Treatment [J]. Research progress. 2010, 30 (5): 32-34. (In Chinese)

[15] Alexander S., Sirotkin, Larisa Yu., Koshkina, Konstantin G, Ippolitov. The BAC-Process for Treatment of Wastewater Containing Non, Ionogenic Synthetic Surfactants [J]. Wat. Res., 2001, 35(13): 3265-3271.

[16] Baoan Zhang, Hongwei Zhang, Xuehua Zhang, Liankai Zhang. Development of the Application of Biological Activated Carbon to Water Treatment [J]. Industrial Water Treatment. 2008, 28(7): 6-8. (In Chinese)

[17] Walker G M, Weateherley L R. Biological Activated Carbon Treatment of Industrial Wastewater in Stirred Tank Reactors [J]. Chemical Engineering Journal, 1999, 75(3):201-206.

[18] Wenjie He, Weiguang Li, Xiaojian Zhang, Tingling Huang, Hongda Han. Novel Technology for Drinking Water Safety [M]. China Architecture & Building press. 2006.

[19] B.Z Wang. The Efficacy and Mechanism of Removal of Organic Substances from Water by Ozone and Activated Carbon. Water Science and Technology 1999, 30(1): 43-47.

[20] Y Wang, J.Hi Qu, R.C Wu, et al. The Electrocatalytic Reduction of Nitrate in Water on Pd/Sn-modified Activated Carbon Fiber Electrode. Water Research. 2006(40): 1224-1232.

[21] P. Wentworth, J. Nieva, T C.akeuchi, et al. Evidence for Ozone Formation in Human Atherosclerotic Arteries. Science. 2003, 302(7): 1053-1056.

[22] Woo Hang Kim, Wataru Nishijima. Micropollutant Removal with Saturated Biological Activated Carbon (BAC) in Ozonation-BAC Process. Water Science and Technology. 1997, 36(12): 283-298.

[23] W. Nishijima, E.G Speitel. Fate of Biodegradable Dissolved Organic Carbon Produced by Ozonation on Biological Activated Carbon. Chemosphere. 2004, 56(2):113-119.

[24] X. Zhao, R.F Hickey, T.C Voice. Long-term Evaluation of Adsorption Capacity in Biological Activated Carbon Fluidized Bed Reactor System. Water Research. 1999, 33(13): 2983-2991.

[25] Anneli Andersson, Patrick Laureent. Impact of Temperature on Nitrification in Biological Activated Carbon (BAC) Filters Used for Drinking Water Treatment. Water Research. 2001, 35(12): 2923-2934.

[26] Xiaorong Wang, Guangping Hao, Wencui Li. Research and application of biological activated carbon for water treatment. Chemical industry and engineering process. 2010, 29(5): 932-937. (In Chinese)

[27] Pengkang Jin, Xu Wang, Jianjun Xu, Xiaochang Wang. Analysis on the process of biological activated carbon bed with the effect of ozonation. Technology of water treatment. 2010, 36(12): 15-18. (In Chinese)

[28] Bozena Seredynska-Sobecka. Removal of Humic Acids by the Ozonation-Biofiltration Process [J]. Desalination, 2006,198 (1/3): 265-273.

[29] Goel S, Hozalski R M, Bouwer E J. Biodegradation of NOM: Effect of NOM Source and Ozone Dose [J]. J AWWA, 1995, 87(1): 90-105.

[30] Greening F. Experience with Ozone Treatment of Water in Switzerland [C]. Andrews G F. Proceedings of 82th Ozone world Congress. Zurich, 1987: 49-54.

[31] Lin Wang, Qifang Luo. Study on Degradation of Immobilized Microorganism on Endocrine Disruptor Di-n-butyl Phthalate [J]. China J Public Health. 2003, 19(11): 1302-1303.

[32] M. Scholz, R.J Martin. Ecological Equilibrium on Biological Activated Carbon [J]. Water. Research. 1997, 31(12): 2959-2968.

[33] R. Narasimmalu, M. Osamu, I. Norifumi, et al. Variation in Microbial Biomass and Community Structure in Sediments of Eutrophicbays as Determined by Phospholipid Ester Linked Fatty Acids [J]. Applied and Environment Microbial.1992, 58(2): 562-571.

[34] G. Collins, A. Woods, M. H Sharon, et al. Microbial Community Structure and Methanogenic Activity during Start-up of Psychrophilic Anaerobic Digesters Treating Synthetic Industrial Wastewaters [J]. FEMS Microbiology Ecology. 2003, 46(1): 159-170.

[35] D.C Ellwood. Adhession of Microorganisms to Surfaces. London: Academic Press. 1979: 62-78.

[36] D.C Savage, M. Fletcher. Bacterial Adhesion: Mechanisms and Physiological Significance [J]. New York: Plenum Press. 1985: 197-206.

[37] J. David, Pernitsky, R. Gordon, et al. Recovery of Attached Bacteria from GAC Fines and Implications for Disinfection Efficacy [J]. Water Research. 1997, 31(3): 385-390.

[38] Y. Takeuchi, Y. Suzuki, K. Mochidzuki, Biological Activated Carbon Treatment of Organic Water Containing Heavy Metal Ions at a High Salt Concentration [J]. Process. SCEJ Kusyu Regional Meeting. 1994: 125-126.

[39] R.Ahmad, A.Amirtharajah. Detachment of Particals during Biofilter Backwashing [J]. Journal AWWA. 1998, 90(12):74-85.

[40] V.D Kooij. Determining the Concentration of Easily Assimilable Organic Carbon in Drinking Water [J]. Journal AWWA. 1992, 74(10): 540-545.

[41] Y. Liu, Q.D Wang. Surface Modification of Biocarrier by Plasma Oxidation-Ferric Ions Coating Technique to Enhance Bacterial Adhesion [J]. Journal of Environmental Science and Healthy. 1996, (3): 869-879.

[42] LeChevallier, M.W Welch. Full-Scale Studies of Factors Related to Coliform Regrowth in Dinking Water [J]. Applied and Environmental Microbial. 1996, 62(7): 2201-2211.

[43] Guangzhi Wang. Study on Biological Stability Control of Dominant Bacteria in Immobilization Biological Activated Carbon [J]. Dissertation for the Doctoral Degree in Engineering, June, 2008, China.

[44] J.G Jacangelo. Selected Processes for Removing NOM: An Overview [J]. Journal AWWA. 1995 (1): 64-77.

[45] D. Susan, Richardson. Disinfection By-products and Other Emerging Contaminants in Drinking Water [J]. Trends in Analytical Chemistry. 2003, 22(10): 255-275.

[46] Jun Ma, Tao Zhang, Zhonglin Chen, Minghao Sui, Xueyuan Li. Pathway of Aqueous Ferric Hydroxide Catalyzed Ozone Decomposition and Ozonation of Trace Nitrobenzene [J]. Environmental science. 2005, 26(2): 78-82.

[47] Laisheng Li, Wanpeng Zhu, Zhonghe Li. The Removal of Difficult Degradation Pollutants by catalytic O₃ catalytic and adsorption technology [J]. China water & waste water. 2002, 18(5): 23-25.

[48] T.A Bellar. The Occurrence of Oraganhalids in Chlorination Drinking Water [J]. J.Am. Water work Assoc. 1974, 66(12): 703.

[49] J.C Kruith, A.J Veer, J.P Hoek. Ozonation and Biological Activated Carbon in Dutch Drinking Water Treatment [J]. Regional Conference on Ozone. Netherland. 1996: 85-101.

[50] He Dao-hong, Gao Nai-yun, Zeng Wen-hui, et al. Biofilm Colonization in Advanced Treatment of Drinking Water Using Biological Activated Carbon Process[J]. Industrial Water & Wastewater, 2006, 37(2): 16-19.(In Chinese)

[51] Bożena Seredyńska-Sobecka. Removal of Humic Acids by the Ozonation–biofiltration Process [J]. Desalination, 2006, 198(3): 265-273.

[52] C.H. Liang, P.C. Chiang, E.E. Chang. Systematic Approach to Quantify Adsorption and Biodegradation Capacities on Biological Activated Carbon following Ozonation [J]. Ozone Science & Engineering, 2003, 25: 351-361.

[53] Qing Tian, Jihua Chen. Application of Bioactivated Carbon (BAC) Process in Water and Wastewater Treatment [J]. Environmental Engineering, 2006, 24(1): 84-87. (In Chinese)

[54] Pengkang Jin, Dewang Jiang, Xiafeng Zhang, Xiaochang Wang. Study on the Characteristics of Microorganism Distribution in Biological Activated Carbon Following Ozonation [J]. Journal of Xi'an University of Architecture and Technology, 2007, 39(6): 829-833.(In Chinese)

[55] Wang Min, Shang Haitao, Hao Chunbo, Luo Peng, Gu Junnong, Diversity and Bacteria Community Structure of Activated Carbon Used in Advanced Drinking Water Treatment, [J].Environmental Science, 2011, 32(5): 1497-1504. (In Chinese)

[56] Liang C H, Chiang P C, Chang E E. Systematic Approach to Quantify Adsorption and Biodegradation Capacities on Biological Activated Carbon Following Ozonation[J]. Ozone Science & Engineering, 2003, 25(5): 351–361.(In Chinese)

[57] He Daohong, Gao Naiyun, Zeng Wenhui, et al. Biofilm Colonization in Advanced Treatment of Drinking Water Using Biological Activated Carbon Process [J]. Industrial Water & Wastewater, 2006, 37(2): 16–19. (In Chinese)

[58] Qiao Tiejun, Zhang Xiaojian. In-situ Substrate Uptake Rate Method for Measuring Microbial Activity [J]. China Water & Wastewater, 2002, 18(7): 80–82.

[59] Bozena S S. Removal of Humic Acids by the Ozonation–biofiltration Process [J]. Desalination, 2006, 198(13): 265–273.

[60] Pengkang Jin, Jianjun Xu, Dewang Jiang, Xiaochang Wang. Quantifying Analysis of the Organic Matter Removal by Adsorption and Biodegradation on Biological Activated Carbon Following Ozonation [J]. Journal of Safety and Environment, 2007, 7(6), 22-25. (In Chinese)

[61] Gao Yanhui, Gu Guowei. Water Pollution Control Engineering [M]. Beijing: Higher Education Press, 1999.

[62] WANG xiaochang, ZHANG chengzhong. Environmental Engineering Science [M]. Beijing: Higher Education Press, 2010.

[63] Qiao Tiejun, Sun Guofen, Study on Microbial Safety of Ozone/Biological Activated Carbon Process [J]. China Water and Wastewater, 2008, 24(5): 31-39.(In Chinese)

[64] Zhang Jinsong, Qiao Tiejun. Water Quality Safety and Safeguard Measures of Ozone-biological Activated Carbon Process [J].Water and Wastewater Treatment, 2009, 35(3): 9-13.(In Chinese)

[65] Qiao Tiejun, Zhang Xihui, Water Quality Safety of Ozonation and Biologically Activated Carbon Process in Application [J].Environment Science, 2009, 30(11): 3311-3315.(In Chinese)

[66] Coleman W.E., Munch J.W., Ringhand H.P., Kay-lor W.H. and Mitchell D.E.,"Ozonation/Postchlori-nation of Humic Acid: A Model for Predicting Drinking Water Disinfection by-Products". Ozone Sic. & Engi-neering, 1992, 14, 51-69.

[67] Kurokawa Y., Maekawa A., Takahashi M. and Hayashi Y.,"Toxicity and Carcinogenicity of Potassium Bromate-A New Renal Carcinogen", Environ. Health Pe-rspect., 1990, 87, 309-335.

[68] X. C. Wang, The Byproducts of Ozonation [J]. Water and Wastewater Treatment, 1998, 24(12): 75-77.(In Chinese)

[69] Zhangbin Pan, Baorui Jia. The Water Quality Safety O₃-BAC Process: A Review [J]. City and Town Water Supply, 2010, 2, 67-70. (In Chinese)

Chapter 5

Methods and Applications of Deoxygenation for the Conversion of Biomass to Petrochemical Products

Duminda A. Gunawardena and Sandun D. Fernando

Additional information is available at the end of the chapter

1. Introduction

Depleting reserves, uncertain economics, and environmental concerns associated with crude oil have prompted an extensive search for alternatives for producing transportation fuels. Biomass has been given close scrutiny due to the emphasis on climate change and the ability of biomass based energy systems to mitigate greenhouse gas (GHG) emissions. Further, its ability to produce fuels and chemicals that are identical to those produced using petroleum resources, makes biomass an important alternative raw material [1, 2]. Biomass can be considered clean, as it contains negligible amounts of sulfur and nitrogen. Consequently, the emissions of SO_2, NO_x are extremely low compared to conventional fossil fuels. The overall CO_2 emission is considered to be neutral, as CO_2 is recycled by the plants through photosynthesis [3]. Moreover, substitution of fossil fuels with biomass-based counterparts could lead to net CO_2 emission reductions [4].

1.1. Biomass production

Biomass encompasses all plant and plant-derived materials as well as animal matter and animal manure. Due to its abundance biomass has to be considered as a vital source of energy to satisfy the global energy demand. Table 1 shows the breakdown of primary energy sources and their contribution to total world energy demand during 1973 and 2009. It is evident that the contribution from biomass is only 10% over this period and in the global scale bioenergy has not yet made any significant impact. However, according to a report published by the United States Department of Agriculture, biomass derived energy has become the highest contributor among all the renewable sources during 2003. According to US statistics 190 million dry tons of biomass is consumed per year, which is equivalent to

3% of current energy consumption [5]. Even though it is hard to conceive that biomass can totally eliminate the use of fossil fuel consumption, it can complement the renewable energy sector together with other sources like wind, solar and geothermal energy.

Primary energy sources	1973	2009
The total world energy demand (Mtoe)	6111	12150
Oil	46.00%	32.80%
Natural gas	16.00%	20.90%
Nuclear	0.90%	5.80%
Hydro	1.80%	2.30%
Biofuels and waste	10.60%	10.20%
Coal / peat	24.60%	27.20%
other	0.10%	0.80%

Table 1. The world's primary energy sources with its contribution. [6]

1.2. Biomass to energy conversions

Biomass has been intimately connected with the everyday life of people since prehistoric times. With the advent of fire by rubbing splinters, humans began to exploit the chemical energy of biomass to produce heat. Ever since, various methods have been developed to convert chemical energy present in biomass to useful heat energy. These conversion methods can be summarized as shown in figure (1).

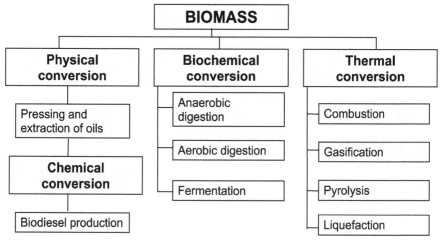

Figure 1. Different categories of biomass conversion

Physical conversion of biomass typically involves pressing the plant (or animal) matter to produce triglyceride oils. Triglycerides cannot be used directly as transport fuels and needs to be processed further. Triglycerides can be converted into a renewable fuel widely known as biodiesel using the transesterification process. This process converts triglyceride in the presence of alcohol to fatty acid alkyl esters .

Biochemical conversion primarily involves using microorganisms or enzymes to breakdown complex chemicals present in biomass into simpler sugars or alcohols. Biomass conversion to alcohols such as ethanol has attracted wide interest in the recent past. The corn to ethanol technology is mature and is commercial in the US. Nevertheless, this technology has created some debate in the context that corn is still a primary food to many around the world. To circumvent this issue, significant strides have been made in the use of lingo-cellulosic biomass as a source for ethanol. In this method enzymes are used to breakdown cellulose to its monomers and subsequently subjected to fermentation under anaerobic conditions using microorganisms.

Thermochemical conversion is another key process that uses heat to induce chemical transformations in the biomass constituents to produce energetically useful intermediate and/or end products. Conversion techniques available under this umbrella can be categorized into four main processes as represented in figure (1). Energy generation by combustion of biomass can be considered as the most archaic [7]. However, increasing demand for transport fuels has led to the development of other processes that involve converting biomass into liquid and gaseous products [8] such as gasification, pyrolysis and liquefaction. Gasification is the conversion of biomass to a mixture of gases called synthesis gas (or syngas) that primarily consists of hydrogen, carbon monoxide, carbon dioxide and methane. During gasification, biomass is heated under an oxygen-lean environment [9]. Synthesis gas can be directly used in an internal combustion engine or can be converted to liquid fuels using a method known as Fisher-Tropsch (FT) synthesis [7, 10]. Fisher-Tropsch process is considered to be quite energy intensive and therefore, is not yet believed to be economical to compete with petroleum fuels. Nevertheless, active research is still in progress to improve the process [11, 12]. Fermentation of synthesis gas to alcohols (primarily ethanol) using microorganisms is also an active area of research [13].

Pyrolysis and liquefaction are two closely related routes targeted towards producing liquids – called bio-oil or bio-crude [14, 15]. Although not universally accepted, the term bio-oil generally refers to the highly oxygenated liquid product that directly exits a pyroloysis reactor. The term bio-crude represents a more deoxygenated liquid product.

Pyrolysis, unlike gasification, takes place in an oxygen-free atmosphere. The most common technique for producing a large volume of condensable fraction is the fast pyrolysis process. Several reactor configurations used for fast pyrolysis include: ablative, entrained flow, rotating core, vacuum pyrolysis, circulating fluidized bed, and deep bubbling fluidized bed reactors [16]. During fast pyrolysis, biomass is heated at a very high heating rate (eg. 10^3–10^4 K/s)[17]. The temperature at which the thermal scission of biomass material such as cellulose and lignin take place is around 450–550 °C. The bio-oil portion is originally in the vapor-phase and obtained by quenching the volatile output. The yields of the condensable-fraction are reported to be as high as 70–80%. According to some kinetics studies conducted for different biomass, the frequency factor for pyrolysis varies between 10^9-10^{11} orders of magnitude. This is an indication of how fast the reaction would occur during a short residence time [18].

Since its chemical complexity, the biomass pyrolysis reaction has been studied by using model compounds like cellulose. A set of possible reactions paths under different heating conditions is presented in figure (2)[19, 20].

The composition of bio-oil is substantially different from crude petroleum due to the presence of high concentrations of oxygenates. Biomass to bio-oil pyrolysis stoichiometry can be represented using an empirical formula as shown in eq.(1).

$$CH_{1.46}O_{0.67} \rightarrow 0.71CH_{1.98}O_{0.76} + 0.21CH_{0.1}O_{0.15} + 0.08CH_{0.44}O_{1.23} \qquad (1)$$
$$\text{Bio oil} \qquad\qquad \text{Char} \qquad\qquad \text{Gas}$$

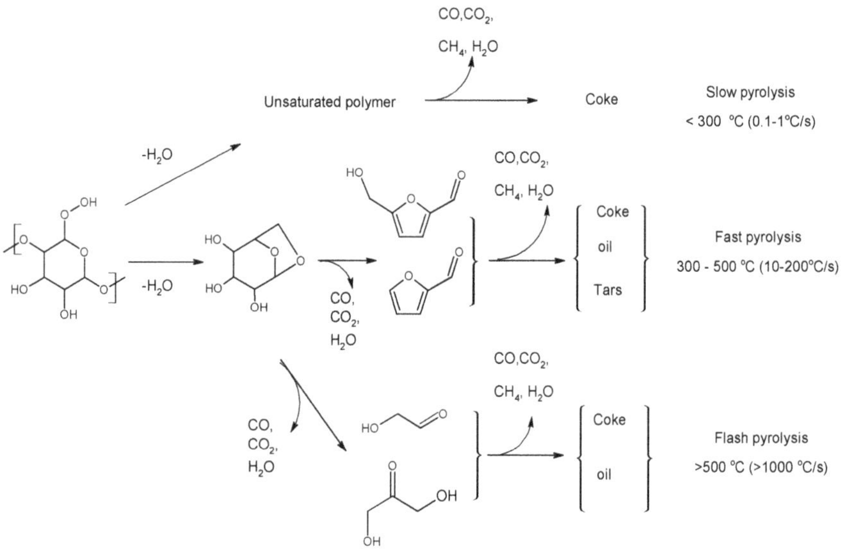

Figure 2. Cellulose pyrolysis with the products at different heating conditions. (Information for scheme was adapted from Huber et al. [20])

Bio-oil properties are highly variable and depend heavily on the type of biomass. High fibrous biomass that contains high amounts of lignin is considered to be the most effective for the production of bio-oil for fuel applications. Further, the oil-yields and composition of bio-oils highly depend on the process (pyrolysis/ liquefaction) used. Fast pyrolysis yields higher amounts of bio-oil compared to slow pyrolysis whereas liquefaction produces low amounts of oxygenated compounds as compared to fast pyrolysis. Oxygenated compounds such as aldehydes, alcohols, ketones, and carboxylic acids can be seen in bio-oils in varying degrees. A typical product distribution of bio-oil is depicted in figure (4). It can be seen that bio-oil contains large amounts of ketones, carboxylic acids and aldehydes [22].

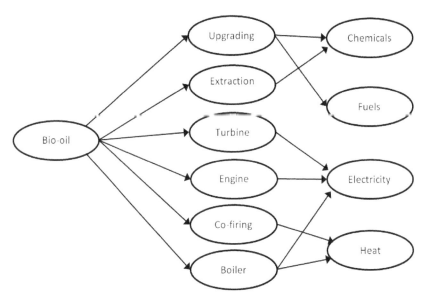

Figure 3. Different industrial applications of bio-oil derived from biomass (In formation adapted from Bridgwater et al. [21].)

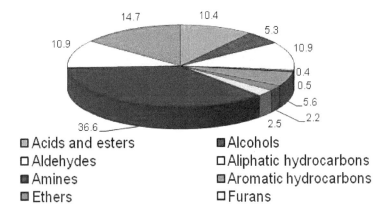

☐ Acids and esters   ■ Alcohols
☐ Aldehydes   ☐ Aliphatic hydrocarbons
■ Amines   ☐ Aromatic hydrocarbons
■ Ethers   ☐ Furans

Figure 4. Relative distribution of chemical compounds in bio-oil (In formation adapted from Adjaye et al. [23].)

Bio-oil has already been tested in furnaces and gas turbines [21], as well as in space heaters and in boilers[24, 25], as represented in figure (3). Although bio-oil has potential as a crude-oil alternative, problems have been reported when bio-oil was used in such applications. These include, blocking filters by high levels of char particulates, high viscosity causing pumping issues, and corrosion from the low pH [26].

Liquefaction is another approach to producing bio-oils. The concept of oil production using biomass in hot water surfaced in the early 1920's. However, a more robust and effective method was not available until Pittsburg Energy Research Center (PERC) in the 70's demonstrated the use of carbon monoxide, steam and sodium carbonate catalyst at Albany Biomass Liquefaction facility in Oregon USA. Detailed information on this method can be found elsewhere [27].

The quality of bio-oil from the liquefaction process is reported to be superior to that obtained from pyrolysis. Liquefaction of biomass using super -critical methods can be considered as one of the more recent techniques under investigation.

Conversion of biomass to a product that is compatible with existing petroleum refinery infrastructure is prudent in several fronts. First, this will allow biomass to be converted into fuels and chemicals that are identical to what we use today (such as gasoline, diesel and jet fuels). So, such fuels could be used in present-day automobiles with no engine modifications. On the other hand, usage of existing fuel production and distribution of infrastructure helps long-term sustainability of biomass-to-fuels technology [28].

The challenge of converting bio-oil into a hydrocarbon fuel has been effective for the removal of functional groups that contain oxygenates (–OH, -COH, -COOH, etc.). The oxygen content of biomass-derived bio-oil is estimated to be 35-40% with a heating value between 16 and 19 MJ/kg [8]. In the recent years, there has been an unprecedented growth of research and developmental efforts related to conversion technologies and the information is scattered. Accordingly, the overall objective of this review is to assimilate this information and compare the status of key deoxygenation technologies.

2. Deoxygenation

The presence of oxygen in cellulose, hemicelluloses and lignin is the primarily reason for biomass derived bio-oil to be highly functionalized. Cellulose and hemicellulose have common structural building blocks which are glucose and xylose respectively. Cellulose which is the most abundant component in terrestrial biomass is a crystalline polymer, and therefore, is quite difficult to chemically transform. Lignin which is abundant in biomass is mainly an amorphous poly aromatic polymer. Because of this complex heterogeneity of biomass, the liquid derived from the pyrolysis or liquefaction process contains a variety of different chemical species which can be roughly estimated to be around 400.

2.1. Importance of deoxygenation

The properties of bio-oil are significantly affected by its chemical composition. Unlike crude petrolium, bio-oil constituents have numerous functional groups as shown in figure (4). The high functionality decreases the stability of the oils and results in polymerization. A direct consequence is the increase of viscosity of bio-oils with time which in turn raises concerns about the feasibility of using bio-oil products as substitutes for petrochemical fuels. In addition to instability, low pH values and the presence of a high water content and ash content are considered problematic [29, 30]. Additionally, the presence of oxygenates and water in the bio-oil reduces the heating value. Consequently, upgrading is necessary to make it useful as a fuel.

As shown in the figure (5) upgrading improves the H/C molar ratio together with increasing the heating value. In contrast, crude petroleum is lean in oxygenated compounds and therefore, most of the chemistries developed to date are based on adding functional groups to increase its activity. As a result, there is only limited knowledge on techniques available to remove functionality from highly oxygenated compounds. However the upgrading of bio-oil by removing of oxygen (deoxygenation) is considered as one of the most intricate challenges we face today in getting bio-oil to a form that is applicable as a fuel intermediate.

As discussed above, chemical species present in biomass derived liquid bio-oil can be categorized into alcohols, aldehyde / ketones, carboxylic acids, phenols and furans. Therefore the upgrading process of bio-oil involves removal of various oxygenated species and can be collectively called as deoxygenation. The deoxygenation process predominantly involves three reaction classes, i.e., dehydration, decarbonilation and decarboxylation.

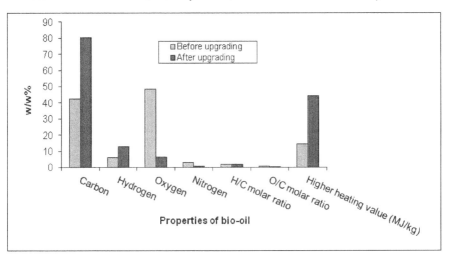

Figure 5. Change of characteristics before and after upgrading bio-oil derived from soybean stalk (Information adapted from Li et al. [8].)

2.2. Reactions involved in deoxygenation

As previously mentioned, deoxygenation involves the removal of functionality of biomass constituents associated with -OH, -COOH, and -C=O. The bond dissociation energies (BDE) for these functional groups are quite high and can be written in descending order as follows:

C-O (1076.5 kJ/mol) > C=O (749 kJ/mol) > C-C (610.0 kJ/mol) > O-H (429.99 kJ/mol) > C-H (338.4 kJ/mol).

Higher BDE for a particular functional group implies that the activation energy required for dissociating this bond (deoxygenation) would also be high. This would dictate rigorous reaction conditions for particularly C-O and C=O bond scissions. The presence of large amount of C=O groups in the pyrolysis products can be related to the higher activation energies required for dissociation of these bonds. However, using appropriate catalysts, these high activation energies could be overcome.

2.2.1. Dehydration

Bio-oil has significant amounts of –OH groups in its components that require dehydration, i.e., removal of oxygen in the form of water, to make hydrocarbons. Dehydration occurs even during HDO process but what is considered here is the spontaneous removal of oxygen as water in the absence of extraneous H_2.

Studies on dehydration as a method to produce motor fuel initiated several decades ago. With the advent of new catalysts, this area grew into new heights. A wide range of studies have been carried out with oxygenates ranging from simple methanol, to polyols like glycerol. The presence of light-molecular-weight alcohols such as methanol and ethanol in bio-oil is less common while phenolic compounds and polyols can be considered to be more abundant. However, studies on light alcohols give good insights into the chemistry and will be discussed in more detail below.

Dehydration of methanol to produce gasoline products such as benzene, toluene and xylene has been reported by different research groups [31-35]. In this regard, different heterogeneous catalysts have been studied. In particular, the molecular sieve ZSM-5 has received great attention for dehydration reactions. It has been widely accepted that Brönsted acid sites of the ZSM-5 catalyst play a crucial role for the dehydration reaction. Acid sites donates protons to the hydroxyl group of the oxygenate as shown in figure (7) to form water instigating dehydration.

Addition of metal oxides onto the catalyst framework was reported to increase the acidity of the support material. As a result, this would enhance dehydration reactions in turn facilitating the formation of higher molecular weight hydrocarbons. For example, methanol conversion to gasoline range hydrocarbons over metal oxides (such as ZnO and CuO) supported on HZSM-5 at temperature 400 °C and 1 atm pressure was reported [36]. The results indicate that pure HZSM-5 produced the lowest yields compared to metal-oxide-promoted catalyst such as CuO/ HZSM-5, CuO /ZnO/HZSM-5, ZnO/HZSM-5. The presence of CuO significantly increased the yields of aromatics. It was further concluded that addition of ZnO over CuO significantly reduced the catalyst deactivation potential [36]. Studies conducted to find the effect of different CuO loadings on methanol conversion indicate that the highest conversion of 97% was obtained when the CuO loading was at 7%. It was observed that further increase of oxide loading decreased methanol conversion. This behavior was attributed to loss of acid sites on the support. A subsequent catalyst deactivation study indicated that the catalyst deactivation increased with the increase in CuO loading. It was concluded that the deactivation of the catalyst occurred mainly due to the deposition of large molecular weight hydrocarbons known as coke blocking the catalyst pores[37].

Dehydration of ethanol has also been studied extensively with other catalysts. A study of the effect of different additives such as ZnO, Ga$_2$O$_3$, Mo$_2$C and Re, on ZSM-5 on deoxygenation of alcohols has provided new insights on renewable aromatic hydrocarbons production [38]. The product selectivities for different catalysts are depicted in figure (8). It is clear that Ga$_2$O$_3$ performed better in terms of selectivity toward benzene, toluene and xylenes found in gasoline.

Brønsted acid site

Lewis acid site

Figure 6. The two form of acid sites that exist in zeolites.

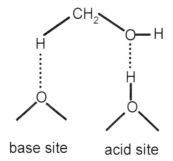

base site   acid site

Figure 7. The proposed adsorption of methanol on ZSM-5.(Information was adapted from Chang et al. [31])

Ethanol dehydration has been further studied in order to make ethylene as the precursor to make ethyl-tetra-butyl-ether (ETBE) [39, 40]. Ethers have gained much attention as a substitute for petroleum diesel. A study on ethanol dehydration to ethylene over dealuminated modernite (DM) and metals such as Zn, Mn, Co, Rh, Ni, Fe and Ag loaded DM has shown that Zn/DM and Zn/Ag/DM gives the highest selectivity to ethylene formation [40]. This indicates that incorporation of single metal or metal combinations onto dealuminated modernite makes the catalyst more selective toward ethylene production. Further , the results suggest that such combinations of metal and support would lower coke formation as high molecular weight compounds were not produced significantly.

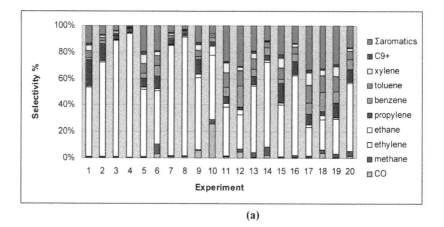

(a)

Experiment No:	Catalyst used	Reaction Temperature -K
1	ZSM-5(80)	773
2	ZSM-5(80)	873
3	ZSM-5(280)	773
4	ZSM-5(280)	873
5	2% Mo₂C/ZSM-5(80)	773
6	2% Mo₂C/ZSM-5(80)	873
7	2% Mo₂C/ZSM-5(280)	773
8	2% Mo₂C/ZSM-5(280)	873
9	2% Re/ZSM-5(80)	773
10	2% Re/ZSM-5(80)	873
11	2% Ga₂O₃/ZSM-5(80)	773
12	2% Ga₂O₃/ZSM-5(80)	873
13	2% ZnO/ZSM-5(80)	773
14	2% ZnO/ZSM-5(80)	873
15	ZSM-5(80) + 2% Mo₂C/ZSM-5(80)	773
16	ZSM-5(80) + 2% Mo₂C/ZSM-5(80)	873
17	ZSM-5(80) + 2% Ga₂O₃/ZSM-5(80)	773
18	ZSM-5(80) + 2% Ga₂O₃/ZSM-5(80)	873
19	ZSM-5(80) + 2% ZnO/ZSM-5(80)	773
20	ZSM-5(80) + 2% ZnO/ZSM-5(80)	873

(b)

Figure 8. (a) Selectivity towards different products from ethanol dehydration. (b) The catalysts and respective reaction conditions for each reaction.(Prepared with data from Barthos et al.[38])

There has been some studies on phenol dehydration [41, 42]. In once such study where HZSM-5 was used it has been observed that the reactivity of phenol and 2-methoxyphenol has been very low and that the catalyst had a greater tendency to form coke [42]. The rate of deactivation of the catalyst by coke formation reduced with increased water formation. In a separate study, lignin derivative guaiacol was attempted to be transformed to phenol at a temperature of 350 °C and 1atm pressure. In this study, first row transition metals (V to Zn) supported on alumina was tested [43]. Results indicate that vanadium oxide on alumina gave the highest yield of phenol. They concluded that vanadium, as an early transition metal, has the oxophilic property and had helped the efficient removal of oxygen from guaiacol in the form of water. Nevertheless, water formed during dehydration has a tendency to adsorb onto acid sites dramatic decreasing the catalyst activity.

2.2.2. Decarboxylation

Bio-oil contains considerable amounts of acids such as acetic, formic, and butyric acid, resulting in low pH values (c.a. pH 2-3). The presence of these acids creates various practical challenges for bio-oil applications. Highly corrosive nature makes bio-oil not suitable for applications with metals and rubber. Further, the presence of acids would increase O/C ratio and make bio-oil more reactive. Accordingly, it is critical to develop chemistries that can deal with carboxylic acids when upgrading bio-oil. Decarboxylation refers to the removal of oxygen in the form of CO_2 from a carboxylated compound and can be given in a general equation as follows (eq.(2)):

$$RCO_2H \rightarrow RH + CO_2 \tag{2}$$

Insights on effective catalysts for removing oxygen in carboxylic acids can be obtained from studies on decarboxylation of model systems that include stearic, palmetic, benzoic, and heptanoic acids. Equations (3) and (4) depict the thermodynamic favorability of the decarboxylation reactions. The negative values for ΔG^o implies that the decarboxylation reactions of acetic and benzoic acids have a significant activation energy barrier to overcome[44].

$$CH_3CO_2H \rightarrow CH_4 + CO_2 \qquad \Delta G^o = -68.5 \ kJ/mol \tag{3}$$

$$C_6H_5CO_2H \rightarrow C_6H_6 + CO_2 \qquad \Delta G^o = -54.3 \ kJ/mol \tag{4}$$

Biodiesel research is an important area to look for information related to decarboxylation. Fatty acids and fatty acid methyl esters from biodiesel industry have been subjected further deoxygenation with the intension of obtaining higher quality liquid fuels [45, 46]. In one such study, Pd has been identified to be an active metal for decarboxylation of fatty acids present in plant oil. In this study, the ability to convert heptanoic acid to octane was investigated using Pd/SiO_2 and Ni/Al_2O_3 catalysts [47-50]. It was reported that 98% acid conversion was obtained with Pd/SiO_2 at 330°C but only 64% was reported by Ni/Al_2O_3 [44].

Pd supported on active carbon has also been tested as the catalyst for decarboxylation of stearic acid at 300°C. The results indicated that the reaction was selective toward n-heptadecane [47]. Further, they claimed that conducting the reaction in the presence of hydrogen increased the rate of decarboxylation. A comparative study performed with thermal and catalytic deoxygenation of stearic acid further proved that catalytic deoxygenation is highly selective toward hydrocarbons. In this study, 5%Pd supported on mesoporous silica, SBA-15, MCM-41and zeolite-Y has been used as the catalyst. It was reported that SBA-15 had a selectivity of 67% for n-pentadecane [48]. The study further revealed that the deoxygenation activity reduces in the order as SBA -15 > MCM-41 > zeolite-Y.

In an analogous study, pure palmitic acid, stearic acid, and a mixture of 59% of palmitic and 40% of stearic acid was deoxygenated over 4 % Pd/C catalyst at 300 oC and 5% H_2 in argon at 17 bar of pressure. The conversion of the catalyst was reported to be over 94% after 180 min of the reaction time with a selectivity of 99% [51]. The kinetic behavior of decarboxylation of ethyl stearate over Pd / C has been investigated with the aim to verify the reaction mechanism. As shown in figure (9), decarboxylation of ethyl stearate proceeded through fatty acid decarboxylation to the desired n-heptadecane. The produced paraffin simultaneously dehydrogenated to unsaturated olefins and aromatics. A kinetic model has been developed based on the proposed reaction network in figure (9) using Langmuir–Hinshelwood mechanism with the assumptions that the surface reaction is rate limiting and the adsorption reaction is rapid compared to surface reaction [52]. The rate equation for the proposed reaction scheme can be represented in a simplified form as shown in eq.(5). According to the rate information, step 6 in figure (9) can be considered as the rate limiting step with the rate constant of 1.45×10^{-12}/min. Both decarboxylation steps which were represented in step 4 and 5 is shown to be the fastest steps in the scheme.

$$r_i = \frac{k_i \cdot C_i}{\left(1 + \sum_i K_i \cdot C_i\right)} \tag{5}$$

(r_i: reaction rate, k_i: lumped reaction rate, C_i: concentration, K_i: equilibrium constant)

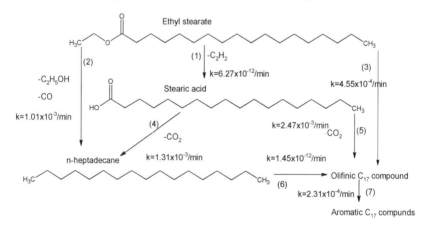

Figure 9. Decarboxylation of elthyl stearate (Information was adapted from Snare et al.[52])

HZSM-5 can be considered as a versatile catalyst that has the ability to do both dehydration and decarboxylation. For example, decarboxylation of methyl esters to hydrocarbon fuels has been studied using methyl octonoate (MO) on HZSM-5 [53]. The catalyst showed strong signs of MO adsorption on to the catalyst surface. This reaction produced significant amounts of C1-C7 hydrocarbon compounds and aromatics. Formation of octonic acid as a primary product indicates that acidic hydrolysis reaction has taken place. However, it was noted that these primary products further undergo conversion into aromatic compounds. The proposed reaction scheme for the MO conversion is presented in figure (10).

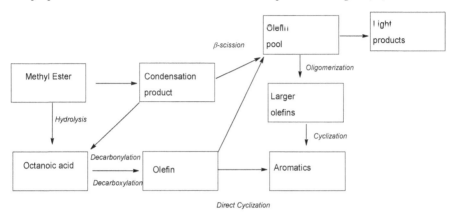

Direct Cyclization

Figure 10. A possible reaction pathway for the deoxygenation of methyl octonoate (Information was adapted from Danuthai et al. [53]).

Rather than complete removal, partial removal of oxygen to aldehydes or ketones would also be useful during upgrading since the latter product(s) can go through HDO pathway relatively easily. Various studies have been conducted in this regard and many have used benzoic acid as the model compound [54-58]. In such a study, different weak base catalysts such as MnO_2, CeO_2, MgO, ZnO, Fe_2O_3, K_2O supported on SiO_2, Al_2O_3, TiO_2 have been tested for upgrading the acid-rich phase of bio-oil through ketonic condensation. The study further evaluated the effect of the presence of water on ketonic condensation of three model components, phenol, *p*-methoxyphenol, and furfural (typically seen in bio-oil). They reported that CeO_2 on Al_2O_3 and TiO_2 had better catalytic activity and tolerance to water. Although the presence of water and phenol did not have a significant impact on the ketonic condensation of acetic acid, the presence of furfural exhibited a strong inhibitory effect on the reaction [54].

Recent studies reported that the best catalysts for the conversion of carboxylic acids to aldehyde /alcohol(s) were Al_2O_3, SiO_2, TiO_2 or MgO supported transition/noble metals such as Pt, Pd, Cu, or Ru. For example, in deoxygenation of methyl stearate and methyl octanoate over alumina-supported Pt [59], 1% $Pt/\gamma-Al_2O_3$ reported to be highly active and selective toward deoxygenation. They reported that 1% Pt/TiO_2 displayed a higher C_8 hydrocarbon selectivity than 1% Pt/Al_2O_3. This was attributed to the presence of larger oxygen vacancies on the TiO_2 support [59]. Results of a similar screening study for the decarboxylation of stearic acid are depicted in the figure (11). It is apparent that Pd, Pt on activated carbon and 5% Ru on MgO resulted in the highest conversion of stearic acids to hydrocarbons.

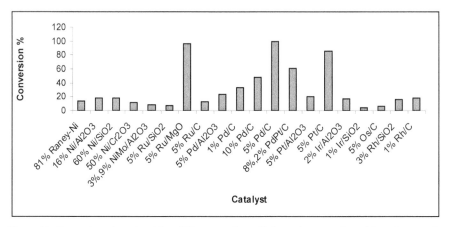

Figure 11. Conversion of stearic acid on different catalysts (Information was adapted from Snare et al. [45].)

2.2.3. Decarbonylation

In general, bio-oil contains significant amounts of aldehydes and ketones c.a. 10.9% and 36.6% respectively. The presence of carbonyl groups in the structure reduces the heating value and stability of bio-oil. Therefore, selective removal of carbonyl group as carbon monoxide as given in eq. (6) is another route to make bio-oil a more favorable fuel intermediate. However, the level of understanding of decarbonylation as a route for the upgrading bio-oil is still quite limited.

$$RCOH \rightarrow RH \; + \; CO \tag{6}$$

According to literature, decarbonylation and decarboxylation are integral reactions in the deoxygenation of carboxylic acids and esters. At times, instead of removing CO_2, removal of CO and H_2O can take place in the deoxygenation step and is considered as decarbonylation. Moreover, product(s) derived by decarbonylation / decarboxylation are not significantly different from those obtained from hydrogenolysis [47].

Decarbonylation usually takes place over supported noble metal catalysts such as Pd/C at elevated temperatures [47]. A study on decarbonylation reaction has been carried out to understand the effect of the presence of Cs on zeolite-X for the deoxygenation of methyl octanoate (MO) as well as the effect of methanol co-feeding with MO [60]. The results indicated that the decarbonylation of MO occurs at a higher rate and for extended periods over CsNaX when co-fed with methanol. The surface analysis revealed that MO strongly adsorbed on basic sites of CsNaX and Cs improved the basicity of the catalyst. It was concluded that not only the basicity of the catalyst but also the polar nature of the zeolite catalyst assisted the decarbonylation process [60].

Deoxygenation of aldehyde, ketone and carboxylic acid containing bio-oil constituents has been studied using model compounds such as acetaldehyde, acetone, butanone, and acetic acid [61]. In this study, HZSM-5 was used as the catalyst. Acetone was considered to undergo a reaction via a mechanism as depicted in figure (12). The results indicated that acetone is less reactive than alcohols and that a higher space velocity was needed to achieve higher conversion into aromatics. A significant increase in coke formation had been observed for both aldehyde and carboxylic acids compared to alcohol.

The ongoing interest in understanding decarbonylation mechanism(s) under the umbrella of organometalic chemistry has resulted in some useful insights. For example, a theoretical and an isotope labeled experimental study of decarbonylation of benzaldehyde and phenyl acetaldehyde on rhodium surface in the presence of bidentate phosphine ligand indicate that decarbonylation mechanism consists of oxidative addition, migratory extrusion, and reductive elimination with migratory extrusion as the rate-determining step [62].

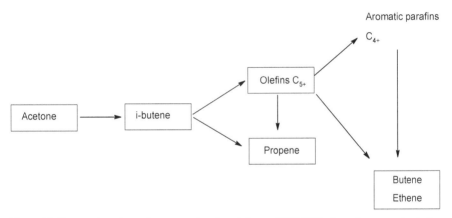

Figure 12. The reaction scheme for acetone decarbonylation on HZSM-5 (In formation was extracted from Gayubo et al. [61].)

Using DFT calculations, it has been deduced that decarbonylation of acetaldehyde is assisted by Co+ as the representative transition metal ion. The study concludes that decarbonylation of acetaldehyde follows four steps, i.e., complexation, C–C activation, aldehyde H-shift and nonreactive dissociation [63].

Furan, C_4H_4O is one of the common oxygenated compounds in the biomass derived bio-oils that has been used to study decarbonylation. Adsorption and desorption steps of furan on pure metal surfaces during the deoxygenation reaction can be found in many publications [64-66]. Some studies on furan decarbonylation has been conducted on different single crystal metal surfaces such as Cu (110), Ag (110) and Pd(111). It was observed that furan absorbs on Cu, Ag and, under mild temperatures, on Pd. Under mild conditions, it was observed that furan desorbs on the metal surface without disrupting the molecule, but, at elevated temperatures, undergoes a deoxygenation reaction [67].

3. Upgrading techniques for fuel production

In any biomass to biofuels conversion process, above mentioned three reaction pathways can be considered to be quite significant. Depending on the process conditions such as temperature, pressure, the resident time and the type of catalyst, the degree to which these reactions would take place may vary. Basically the major processes that are exploited during deoxygenation pathways are secondary cracking, fast catalytic cracking and hydrodeoxygenation.

3.1. Secondary pyrolysis

This is the simplest process with no catalyst involved for the deoxygenation of biomass oxygenates and occurs with fairly low efficiencies. The concept behind this process is simply to route the pyrolysis vapor through a second reactor which is maintained at a high temperature. The thermal cracking reaction initiated in the secondary reactor would remove oxygen as H_2O, CO_2 and CO. Due to the thermodynamic nature of this reaction, increased resident times would favor formation of thermodynamically more stable species such as CO_2 and CO. The significant drawback of this method is the higher tendency of losing carbon in the forms of CO_2 and CO.

3.2. Catalytic upgrading

Catalytic upgrading process is conducted in a number of different ways. Most commonly practiced method would be injection of bio-oil into a tubular reactor packed with a catalyst capable of deoxygenating the substrates. Due to the inefficacies associated with condensing and reheating processes, a reactor capable of accepting a direct feed of pyrolysis vapor into its catalytic chamber has become more popular. Numerous forms of Zeolites are known to be effective for deoxygenation[68] of which ZSM-5 is widely touted to be the most effective catalyst for deoxygenation (via decarboxylation, decarbonylation or dehydration pathways). ZSM-5 is a molecular sieve with 5.5 Å pore channels. This pore structure is responsible for the high selectivity of ZSM-5 toward aromatic hydrocarbons. Deoxygenation reaction on ZSM-5 is believed to be catalyzed at the Bronsted acid sites and its structure is depicted in fig.6.

3.3. Hydrodeoxygenation (HDO)

Due to the hydrogen deficient nature of biomass (C:H<1), catalytic upgrading of bio-oil often leads to deoxygenation via decarboxylation and/or decarbonylation routes leading to losing of precious carbon assimilated during photosynthesis. Analogously, dehydration, in a hydrogen-lean environment leads to formation of large unsaturated compounds commonly known as coke. In order to circumvent these issues, extra hydrogen is supplied to the reactor - and this process is called hydrodeoxygenation.

So far the most reliable and extensively studied method for deep deoxygenation is hydrodeoxygenation which involves gaseous hydrogen and heterogeneous catalyst such as sulfided NiMo, CoMo supported on alumina[69] [70].The idea of using hydrogen to upgrade bio-oil originates from the use of hydrogen in the petrochemical industry. Hydroprocessing is a crucial step in petroleum refining process that basically involves five types of reaction classes: hydrodenitrogenation (HDN), hydrodesulfurization (HDS), hydrodeoxygenation (HDO), hydrodemthylation (HDM) and hydrogenation (HYD)[70-72].

The process where oxygen in the feed is removed via dehydration using gaseous hydrogen is called hydrodeoxygenation (HDO). In a typical hydro processing process, the order of these reaction classes are HDS>HDO>HDN. This is because in a conventional petroleum feed, the sulfur and nitrogen content is significantly higher compared to oxygen. Therefore, HDO chemistry has received only little attention in petrochemical refining [27]. Hydrotreatment of crude petroleum is challenging for the catalyst due to the presence of sulfur and nitrogen in the feed in significant amounts. The products of hydrotreatment such as water, ammonia, hydrogen sulfide has been reported to poison hydrotreating catalysts [73]. Nevertheless, since bio-oil contains less sulfur and nitrogen, HDO would be a better fit for bio-oil upgrading. The presence of significant amounts of oxygen and C=C compounds in bio-oil increases chances of simultaneous occurrence of HDO as well as the hydrogenation reactions. More negative Gibbs free energies of deoxygenation reactions compared to hydrogenation reactions implies that   deoxygenation is more favorable. However, saturation of aromatic rings is not desirable as it consumes large amounts of hydrogen and reduces the octane number of the fuel reducing the fuel quality.

Much of the studies on HDO are based on catalysts such as Co-Mo, Ni-Mo, Ni-W, Ni, Co, and Pd. A catalyst, to be effective for HDO, should ideally perform two tasks, i.e., activating the dihydrogen molecule as well as activating the oxygen group of the compound. The oxygen group activation usually occurs on the transition metal oxides such as Mo, W, Co, Mn, Zr, Ce, Y, Sr and La while the activation of hydrogen is known to happen on noble metals such as Pt, Pd, and Rh [74].

Studies on HDO chemistry has mostly been done using model compound such as phenol, cresole, guaiacol, napthol etc. [75, 76] which are abundantly present in bio-oil. Some of these model compounds and their proposed reaction pathways are shown in Table (2).

Most of the earlier work in HDO used sulfided forms of Mo as the active element and Co or Ni as promoters on $\gamma.Al_2O_3$ [27, 70, 79]. However, sulfide catalyst would not work for bio-oil since the feed does not contain significant amounts of sulfur. In the HDO process, a sulfide catalyst would soon be deactivated if an external sulfur source is not provided [80]. Nevertheless, the oxygen content of the feed is said to have a negative effect on the sulphide structure resulting in losses such as the catalyst deactivation and changes in product distribution. In-depth studies on Co (or Ni)-promoted MoS_2 catalysts have revealed that the edges on MoS_2 is also important in terms of catalyst activity since they are dominated by promoter atoms in the so-called Co–Mo–S structures [81]. The studies have revealed that poly-condensation products formed have shortened the life by deactivation. Alumina in this regard is quite susceptible to deactivation by coke formation. Therefore, investigations on new catalysts that do not require sulfidation and supports such as activated carbon that are tolerant to deactivation are needed[82].

Oxygenate compound	Proposed hydro-deoxygenation reaction	Ref.
OH Phenol		[77]
OH O—CH$_3$ Guaicol		[78]

Table 2. Common reaction pathways proposed for HDO reactions. (Reaction schemes were extracted from Senol et al [77]and Laurent et al.[78])

The suitability of Ni as a catalyst to activate the dihydrogen molecule in HDO has been reported by several groups[79, 83]. In comparison to noble metals, the use of Ni is extremely economical especially for large scale applications. For example, the gas-phase hydrodeoxygenation of a series of aromatic alcohols that include aldehydes and acids has been reported with Ni/SiO$_2$. This study analyzed kinetic effects of the gas phase hydrogenolysis of –CH$_2$OH, –CHO and –COOH groups attached on to aromatic ring structures in the presence of Ni/SiO$_2$ [79]. They concluded that the oxygenated aromatics get weakly adsorbed on the catalyst and the surface mobility facilitates reaction with adsorbed hydrogen atoms. Further, the adsorption reactions of H$_2$ and aromatic species were considered to be a competitive adsorption.

In a separate study, HDO of model compound anisole using Ni-Cu on Al$_2$O$_3$, CeO$_2$ and ZrO$_2$ supports found that Ni–Cu supported on CeO$_2$ and Al$_2$O$_3$ was the most active catalyst in comparison to pure Ni catalyst[74]. The significance of this study was that the catalysts tested were not in the sulfided form and therefore would be highly suitable for bio-oil-type application(s).

The need for hydrogen in HDO process has always been a point of concern due to high expenses associated with hydrogenations. A study in this regard attempted using hydrogen generated in situ for performing HDO. A study conducted using Pt on TiO_2, CeO_2, and ZrO_2 supports showed that the oxygenates undergo dehydrogenation and subsequently HDO using the produced H_2. The catalysts tested were Pt/CeO_2, $Pt/CeZrO_2$, Pt/TiO_2, Pt/ZrO_2, Pt/SiO_2-Al_2O_3, and Pt/Al_2O_3. Of all catalysts tested, Pt on Al_2O_3 showed the highest activity with a reduction of oxygen content from 41.4 wt% to 2.8 wt% after upgrading [84].

It has also been shown that Pd on supports such as carbon, Al_2O_3, ZSM-5, MCM 41 [76, 85, 86] are active for HDO. A study conducted with benzophenone with 5% Pd on active carbon and on ZSM-5 supports proved to be very active toward hydrogenation as opposed to supports such as Al_2O_3, and MCM 41. However, the HDO of benzophenone was significantly higher with Pd on supports like active carbon and acid zeolytes. Furthermore, it was concluded that the acidity of the zeolite support affects the HDO reaction [76].

Co processing of bio-oil with straight run gas oil (SRGO) can be considered to have more practical significance. The concept behind this method is the simultaneous use of HDS process with the HDO of biocrude [69, 87-89]. The model compound guaiacol (5000 ppm) representing oxygenates in bio-oil has been used together with SRGO (containing13,500 ppm of S) in a trickle bed reactor. At low temperature and low space velocities, a decrease in the HDS reaction has been observed with $CoMo/Al_2O_3$. The possible explanation is the competition of intermediate phenol with the sulfur containing molecules for adsorption on hydrogenation/ hydrogenolysis sites. At higher temperatures (above 380⁰C), the HDO of guaiacol was observed along with HDS taken place without further inhibition [88] .

Although the role of sulfur on HDS is well understood, the effect of sulfur on HDO is not yet well explained. Certain studies indicate that the presence of H_2S has an inhibitory effect on the HDO while other studies support maintaining the sulfidasation level of the catalyst [90-92]. For example, the effect of using a sulfiding agent, H_2S has been studied with model compounds like phenol and anisole during hydrotreatment. The results conclude that the presence of H_2S decreases the HDO activity of the sulfided $CoMo/\gamma$-Al_2O_3 catalyst and the product yield depends on the concentration of H_2S [80]. A similar study conducted using H_2S for HDO of phenol on Co-Mo and Ni-Mo arrived at the same conclusion. However, quite interestingly, the presence of H_2S during the HDO of aliphatic oxygenates has shown a promoting effect. The reason for this observation is that the sulfiding agent , H_2S, enhances the acid catalyzed reactions of aliphatic oxygenates. However, direct hydrogenolysis reaction of phenol is suppressed due to competitive adsorption of both phenol and H_2S [77, 93].

4. Concluding remarks

During pyrolysis or liquefaction, biomass undergoes a series of complex conversion processes producing bio-oil. Bio-oils are highly diverse in its composition and has a spectrum of oxygenates. Due to the presence of a collective of different oxygenated compounds, an upgrading process to remove oxygen would likely involve a series of reactions. This review introduces several possibilities for upgrading bio-oil including dehydration, decarboxylation and decarbonylation. A summary of the best catalyst contenders for each reaction class is given in the Table 3.

Reaction class	Reaction details	Catalyst	Best performance
Dehydration	Methanol conversion to gasoline range products	HZSM-5, ZnO/HZSM-5, CuO/HZSM-5	CuO/HZSM-5 (7% loading of CuO is said to be best)
	Ehtanol conversion to hydrocarbon	HZSM-5 ZnO/ ZSM-5 Ga_2O_3/ZSM-5 Mo_2C/ ZSM-5 Re/ ZSM-5	Ga_2O_3/ZSM-5
	Ehtanol to ethylene	dealuminated modernite (DM) Zn/DM Mn/DM Co/DM Rh/DM Ni/DM Fe/ DM Ag/DM	Zn on dealuminated modernite
Decarboxylation	Conversion of heptanoic acid to octane	Pd/SiO_2 Ni/Al_2O_3	Pd/SiO_2
	Deoxygenation of stearic acid	5%Pd supported on messoporous silica - SBA15, -MCM41 zeolite-Y	5%Pd -SBA15
	Ketonic condensation of acetic acid	MnO_2, CeO_2, MgO, ZnO, Fe_2O_3, K_2O supported on SiO_2, Al_2O_3, TiO_2	CeO_2 supported on Al_2O_3, or on TiO_2
Decarbonylation	Deoxygenation of methyl octanoate	Cs on NaX zeolite	Cs on NaX zeolite
	Deoxygenation of acetaldehyde, acetone, butanone, and acetic acid	HZSM-5	HZSM-5

Table 3. Summary of best performing catalyst for each reaction class.

Hydrodeoxygenation has been the most frequently studied and one of the most reliable methods that could be used for deoxygenation of oxygenates. However, the drawbacks of hydrodeoxygenation include: the need for continuous replenishment of sulfur (for sulfided catalysts), and the non-selective deoxygenation of all bio-oil chemical moieties resulting in a spectrum of short-chained and long-chained hydrocarbons that are less useful as liquid fuels. The need of hydrogen makes resulting biofuels less competitive with existing petroleum fuels. More research is needed on development of effective non sulfided hydrodeoxygenation catalysts. Preliminary investigations with Ni-Cu on CeO_2 or ZrO_2 may provide new directions on this front. Further research should be directed on identifying materials and processes will allay the need for using direct hydrogen.

According to the analysis, metal promoted or unpromoted HZSM-5 proves to be one of the most versatile catalysts that is capable of catalyzing all three deoxygenation reaction classes, i.e., dehydration, decarboxylation and decarbonylation. Nevertheless, the nanopore structure of HZSM-5 (~5.4-5.5 Å), while assisting size selectivity for gasoline range hydrocarbons, also promotes pore blockages resulting in rapid catalyst deactivation. Developing catalysts of the same class, with larger and consistent pore structure and/or higher activity/functionality should be closely looked at.

Author details

Duminda A. Gunawardena and Sandun D. Fernando
Department of Biological and Agricultural Engineering, Texas A&M University, College Station, Texas, USA

5. References

[1] Demirbas, A., Biofuels sources, biofuel policy, biofuel economy and global biofuel projections. Energy Conversion and Management 2008. 49: p. 2106–2116.

[2] Demirbas, M.F. and M. Balat, Recent advances on the production and utilization trends of bio-fuels: A global perspective. Energy Conversion and Management 2006. 47: p. 2371–2381.

[3] Bridgwater, A.V., D. Meier, and D. Radlein, An overview of fast pyrolysis of biomass. Organic Geochemistry 1999. 30: p. 1479-1493.

[4] Bhattacharya, S.C., et al., Sustainable biomass production for energy in selected Asian countries. Biomass and Bioenergy 2003. 25(5): p. 471 – 482.

[5] Perlack, R.D., et al., Biomass as a feedstock for a bioenergy and bioproducts industry: The technical feasibility of a billion-ton annual supply., O.R.N. Laboratory, Editor 2005, U.S. Department of Energy U.S. Department of Agriculture.

[6] Association, I.E., Key World Energy Statistics 2011, 2011: Paris.

[7] Hayes, D.J., An examination of biorefining processes, catalysts and challenges. Catalysis Today, 2009. 145 (1-2): p. 138-153.

[8]   Li, J., L. Wu, and Z. Yang, Analysis and upgrading of bio-petroleum from biomass by direct deoxy-liquefaction. Journal of Analytical and Applied Pyrolysis, 2008. 81(2): p. 199–204.

[9]   Reed, T.B. and A. Das, Handbook of Biomass Downdraft Gasifier Engine System, 1988, Solar Energy Research Institue: Golden, Colorado.

[10] Klerk, A.d., Hydroprocessing peculiarities of Fischer-Tropsch syncrude. Catalysis Today, 2008. 130(2-4): p. 439.

[11] Maitlis, P.M., Fischer–Tropsch, organometallics, and other friends. Journal of Organometallic Chemistry 2004. 689: p. 4366–4374.

[12] Lamprecht, D., Hydrogenation of Fischer-Tropsch Synthetic Crude. Energy & Fuels, 2007. 21(5): p. 2509-2513.

[13] Henstra, A.M., et al., Microbiology of synthesis gas fermentation for biofuel production. Current Opinion in Biotechnology, 2007. 18(3): p. 200-206.

[14] Kleinert, M., J.R. Gasson, and T. Barth, Optimizing solvolysis conditions for integrated depolymerisation and hydrodeoxygenation of lignin to produce liquid biofuel. Journal of Analytical and Applied Pyrolysis, 2009. 85: p. 108–117.

[15] Li, H., et al., Liquefaction of rice straw in sub- and supercritical 1,4-dioxane–water mixture. Fuel Processing Technology 2009. 90: p. 657–663.

[16] Mullaney, H., et al., Technical, Environmental and Economic Feasibility of Bio-Oil in New Hampshire's North Country, 2002, University of New Hampshire.

[17] Zhang, Q., et al., Upgrading Bio-oil over Different Solid Catalysts. Energy & Fuels, 2006. 20(6): p. 2717-2720.

[18] Senneca, O., Kinetics of pyrolysis, combustion and gasification of three biomass fuels. Fuel Processing Technology 2007. 88: p. 87–97.

[19] Shafizadeh, F., Introduction to Pyrolysis of biomass. Journal of Analytical and Applied Pyrolysis,, 1982. 3: p. 283-305.

[20] Lin, Y.-C., et al., Kinetics and Mechanism of Cellulose Pyrolysis. The Journal of Physical Chemistry C, 2009. 113(46): p. 20097-20107.

[21] Bridgwater, T., Applications for Utilisation of Liquids Produced by Fast Pyrolysis of Biomass. Biomass and Bioenergy, 2007. 31(8): p. I-VII.

[22] Fernando, S., et al., Biorefineries:Current Status, Challenges, and Future Direction. Energy & Fuels, 2006. 20(4): p. 1727-1737.

[23] Adjaye, J.D. and N.N. Bakhshi, Production of hydrocarbons by catalytic upgrading of a fast pyrolysis bio-oil. Part I: Conversion over various catalysts. Fuel Processing Technology 1995. 45: p. 161-183.

[24] Bridgwater, T., Biomass for energy. Journal of the Science of Food and Agriculture, 2006. 86(12): p. 1755-1768.

[25] Hoekstra, E., et al., Fast Pyrolysis of Biomass in a Fluidized Bed Reactor: In Situ Filtering of the Vapors. Industrial & Engineering Chemistry Research, 2009. 48: p. 4744–4756.

[26] Bridgwater, A.V., Production of high grade fuels and chemicals from catalytic pyrolysis of biomass. Catalysis Today, 1996. 29(1-4): p. 285.

[27] Elliott, D.C., Historical Developments in Hydroprocessing Bio-oils. Energy & Fuels 2007. 3(21): p. 1792-1815.

[28] Huber, G.W. and A. Corma, Synergies between Bio- and Oil Refineries for the Production of Fuels from Biomass. Angewandte Chemie Int. Ed. , 2007. 46(38): p. 7184 – 7201.

[29] Radovanovic, M., et al., Some remarks on the viscosity measurement of pyrolysis liquids. Biomass and Bioenergy 2000. 18: p. 209 222.

[30] Qi, Z., et al., Review of biomass pyrolysis oil properties and upgrading research. Energy Conversion and Management 2007. 48: p. 87–92.

[31] Chang, C.D. and A.J. Silvestri, The Conversion of Methanol and Other O-Compounds to Hydrocarbons over Zeolite Catalysts. Journal of catalysis 1977. 47: p. 249-259.

[32] Dass, D.V. and A.L. Odell, A comparative study of the conversion of ethanol and of ethylene over the "Mobil" zeolite catalyst, H-ZSM-5. An application of the Benzene Sequestration Test. Canadian journal of chemistry, 1989. 67: p. 1732-1734.

[33] Barthos, R., et al., Aromatization of methanol and methylation of benzene over Mo2C/ZSM-5 catalysts. Journal of Catalysis 2007. 247(368-378).

[34] Barrett, P.A., et al., Rational design of a solid acid catalyst for the conversion of methanol to light alkenes: synthesis, structure and performance of DAF-4. Chemical communication, 1996: p. 2001-2002.

[35] Chang, C.D., W.H. Lang, and R.L. Smith, The Conversion of Methanol and Other O-Compounds to Hydrocarbons over Zeolite Catalysts II. Pressure Effects. Journal of Catalysis, 1979. 56: p. 169-173.

[36] Zaidi, H.A. and K.K. Pant, Catalytic conversion of methanol to gasoline range hydrocarbons Catalysis Today, 2004. 96(3): p. 155-160.

[37] Zaidi, H.A. and K.K. Pant, Transformation of Methanol to Gasoline Range Hydrocarbons Using HZSM-5 Catalysts Impregnated with Copper Oxide. Korean Journal of Chemical Engineering, 2005. 22(3): p. 353-357.

[38] Barthos, R., A. Sze´chenyi, and F. Solymosi, Decomposition and Aromatization of Ethanol on ZSM-Based Catalysts. Journal of Physical Chemistry- B 2006. 110: p. 21816-21825.

[39] Phillips, C.B. and R. Datta, Production of Ethylene from Hydrous Ethanol on H-ZSM-5 under Mild Conditions. Industrial Engineering Chemmical Reservs 1997. 36: p. 4466-4475.

[40] Arenamnarta, S. and W. Trakarnprukb, Ethanol Conversion to Ethylene Using Metal-Mordenite Catalysts. International Journal of Applied Science and Engineering, 2006. 4(1): p. 21-32.

[41] Sajiki, H., et al., Pd/C-Catalyzed Deoxygenation of Phenol Derivatives Using Mg Metal and MeOH in the Presence of NH4OAc. Organic Letters 2006. 8(5): p. 987-990.

[42] Gayubo, A.G., et al., Transformation of Oxygenate Components of Biomass Pyrolysis Oil on a HZSM-5 Zeolite. I. Alcohols and Phenols. Industrial & Engineering Chemistry Research, 2004. 43: p. 2610-2618.

[43] Filley, J. and C. Roth, Vanadium catalyzed guaiacol deoxygenation. Journal of Molecular Catalysis A: Chemical 1999. 139: p. 245-252.

[44] Maier, W., et al., Gas phase decarboxylation of carboxylic acids. Chemische Berichte 1982. 115(2): p. 808-812.

[45] Snare, M., et al., Heterogeneous Catalytic Deoxygenation of Stearic Acid for Production of Biodiesel. Industrial & Engineering Chemistry Research, 2006. 45: p. 5708-5715.

[46] Kubickova, I., et al., Hydrocarbons for diesel fuel via decarboxylation of vegetable oils. Catalysis Today, 2005. 106 p. 197–200.

[47] Arvela, P.M., et al., Catalytic Deoxygenation of Fatty Acids and Their Derivatives. Energy & Fuels 2007. 21: p. 30-40.

[48] Lestari, S., J. Beltramini, and G.Q.M. Lu. Catalytic deoxygenation of stearic acid over palladium supported on acid modified mesoporous silica. in Zeolites and Related Materials: Trends, Targets and Challenges Proceedings of 4th International FEZA Conference. 2008.

[49] Snare, M., et al., Catalytic deoxygenation of unsaturated renewable feedstocks for production of diesel fuel hydrocarbons. Fuel 2008. 87: p. 933–945.

[50] Simakova, I., et al., Deoxygenation of palmitic and stearic acid over supported Pd catalysts: Effect of metal dispersion. Applied Catalysis A: General, 2009. 355 p. 100-108.

[51] Lestari, S., et al., Catalytic Deoxygenation of Stearic Acid and Palmitic Acid in Semibatch Mode. Catalytic Letters, 2009. 130: p. 48-51.

[52] Snare, M., et al., Production of diesel fuel from renewable feeds: Kinetics of ethyl stearate decarboxylation. Chemical Engineering Journal 2007. 134: p. 29–34.

[53] Danuthai, T., et al., Conversion of methylesters to hydrocarbons over an H-ZSM5 zeolite catalyst. Applied Catalysis A: General 2009. 361: p. 99–105.

[54] Deng, L., Y. Fu, and Q.-X. Guo, Upgraded Acidic Components of Bio-oil through Catalytic Ketonic Condensation. Energy Fuels, 2009. 23(1): p. 564-568.

[55] Sakata, Y., C.A.v. Tol-Koutstaal, and V. Ponecz, Selectivity Problems in the Catalytic Deoxygenation of Benzoic Acid. Journal of Catalysis 1997. 169: p. 13–21.

[56] Sakata, Y. and V. Ponec, Reduction of benzoic acid on CeO2 and, the effect of additives. Applied Catalysis A: General 1998. 166 p. 173-184.

[57] Lange, M.W.d., J.G.v. Ommen, and L. Lefferts, Deoxygenation of benzoic acid on metal oxides 2. Formation of byproducts. Applied Catalysis A: General 2002. 231 p. 17–26.

[58] Dury, F. and E.M. Gaigneaux, The deoxygenation of benzoic acid as a probe reaction to determine the impact of superficial oxygen vacancies (isolated or twin) on the oxidation performances of Mo-based oxide catalysts. Catalysis Today, 2006. 117(1-3): p. 46.

[59] Do, P.T., et al., Catalytic Deoxygenation of Methyl-Octanoate and Methyl-Stearate on Pt/Al2O3. Catalysis Letters, 2009.

[60] Sooknoi, T., et al., Deoxygenation of methylesters over CsNaX. Journal of Catalysis 2008. 258: p. 199-209.

[61] Gayubo, A.G., et al., Transformation of Oxygenate Components of Biomass Pyrolysis Oil on a HZSM-5 Zeolite. II. Aldehydes, Ketones, and Acids. Industrial & Engineering Chemistry Research, 2004. 43(11): p. 2619-2626.

[62] Fristrup, P., et al., The Mechanism for the Rhodium-Catalyzed Decarbonylation of Aldehydes: A Combined Experimental and Theoretical. J. AM. CHEM. SOC. 2008, 130, , 2008. 130(45): p. 5206–5215.

[63] Zhao, L., et al., Does the Co+-assisted decarbonylation of acetaldehyde occur via C–C or C–H activation? A theoretical investigation using density functional theory. Chemical Physics Letters 2005. 414: p. 28–33.

[64] Caldwell, T.E. and D.P. Land, J. Phys. Chem. B 1999. 103: p. 7869.

[65] Caldwell, T.E., I.M. Abdelrehim, and D.P. Land, Furan Decomposes on Pd(111) at 300 K To Form H and CO plus C3H3, Which Can Dimerize to Benzene at 350 K. J. Am. Chem. Soc., 1996. 118(1): p. 907-908.

[66] Ormerod, R.M., et al., Chemisorption and reactivity of furan on Pd{111} Surface Science, 1996. 360(1-3): p. 1-9.

[67] Knight, M.J., et al., The adsorption structure of furan on Pd(111). Surface Science 2008. 602(14): p. 2524–2531.

[68] Putun, E., B.B. Uzun, and A.E. Putun, Rapid Pyrolysis of Olive Residue. 2. Effect of Catalytic Upgrading of Pyrolysis Vapors in a Two-Stage Fixed-Bed Reactor. Energy & Fuels 2009. 23: p. 2248–2258.

[69] Pinheiro, A., et al., Impact of Oxygenated Compounds from Lignocellulosic Biomass Pyrolysis Oils on Gas Oil Hydrotreatment. Energy & Fuels 2009. 23: p. 1007–1014.

[70] Furimsky, E., Catalytic hydrodeoxygenation. Applied Catalysis A: General 2000. 199: p. 147–190.

[71] Yang, Y.Q., C.T. Tye, and K.J. Smith, Influence of MoS2 catalyst morphology on the hydrodeoxygenation of phenols. Catalysis Communications 2008. 9: p. 1364–1368.

[72] Kim, S.C. and F.E. Massoth, Kinetics of the Hydrodenitrogenation of Indole. Industrial & Engineering Chemistry Research, 2000. 39(6): p. 1705-1712.

[73] Laurent, E. and B. Delmon, Study of the hydrodeoxygenation of carbonyl, carboxylic and guaiacyl groups over sulfided CoMo/[gamma]-Al2O3 and NiMo/[gamma]-Al2O3 catalyst: II. Influence of water, ammonia and hydrogen sulfide. Applied Catalysis A: General, 1994. 109(1): p. 97-115.

[74] Yakovlev, V.A., et al., Development of new catalytic systems for upgraded bio-fuels production from bio-crude-oil and biodiesel. Catalysis Today, 2009. 144(3-4): p. 362.

[75] Odebunmi, E.O. and D.F. Ollis, Catalytic Hydrodeoxygenation I. Conversions of o-, p-, and m-Cresols. Journal of Catalysis   1983. 80: p. 56-64.

[76] Bejblova, M., et al., Hydrodeoxygenation of benzophenone on Pd catalysts. Applied Catalysis A: General 2005. 296: p. 169–175.

[77] Senol, O.I., et al., Effect of hydrogen sulphide on the hydrodeoxygenation of aromatic and aliphatic oxygenates on sulphided catalysts. Journal of Molecular Catalysis A: Chemical 2007. 277: p. 107-112.

[78] Laurent, E. and B. Delmon, Study of the hydrodeoxygenation of carbonyl, carboxylic and guaiacyl groups over sulfided CoMo/[gamma]-Al2O3 and NiMo/[gamma]-Al2O3 catalysts: I. Catalytic reaction schemes. Applied Catalysis A: General, 1994. 109(1): p. 77.

[79] Keane, M.A. and R. Larsson, Application of the Selective Energy Transfer Model to Account for an Isokinetic Response in the Gas Phase Reductive Cleavage of Hydroxyl, Carbonyl and Carboxyl Groups from Benzene Over Nickel/Silica. Catalysis Letters, 2009. 129(1-2): p. 93 -103.

[80] Viljava, T.R., R.S. Komulainen, and A.O.I. Krause, Effect of H2S on the stability of CoMo/Al2O3 catalysts during hydrodeoxygenation. Catalysis Today 2000. 60: p. 83-92.

[81] Lauritsen, J.V., et al., Atomic-scale insight into structure and morphology changes of MoS2 nanoclusters in hydrotreating catalysts. Journal of Catalysis, 2004. 221: p. 510–522.

[82] Puente, G.d.l., et al., Effects of Support Surface Chemistry in Hydrodeoxygenation Reactions over CoMo/Activated Carbon Sulfided Catalysts†. Langmuir 1999. 15: p. 5800-5806.

[83] Shin, E.-J. and M.A. Keane, Gas-Phase Hydrogenation/Hydrogenolysis of Phenol over Supported Nickel Catalysts. Industrial & Engineering Chemistry Research, 2000. 39(4): p. 883-892.

[84] Fiska, C.A., et al., Bio-oil upgrading over platinum catalysts using in situ generated hydrogen. Applied Catalysis A: General, 2009.

[85] Mori, A., et al., Palladium on carbon-diethylamine-mediated hydrodeoxygenation of phenol derivatives under mild conditions. Tetrahedron Letters, 2007. 63: p. 1270–1280.

[86] Procha´zkova, D., et al., Hydrodeoxygenation of aldehydes catalyzed by supported palladium catalysts. Applied Catalysis A: General 2007. 332: p. 56-64.

[87] Sebos, I., et al., Catalytic hydroprocessing of cottonseed oil in petroleum diesel mixtures for production of renewable diesel. Fuel 2009. 88: p. 145–149.

[88] Bui, V.N., et al., Co-processing of pyrolisis bio oils and gas oil for new generation of bio-fuels: Hydrodeoxygenation of guaïacol and SRGO mixed feed. Catalysis Today, 2009. 143(1-2): p. 172.

[89] Donnis, B., et al., Hydroprocessing of Bio-Oils and Oxygenates to Hydrocarbons. Understanding the Reaction Routes. Topics in Catalysis, 2009. 52(3): p. 229-240.

[90] Romero, Y., et al., Hydrodeoxygenation of benzofuran and its oxygenated derivatives (2,3-dihydrobenzofuran and 2-ethylphenol) over NiMoP/Al2O3 catalyst. Applied Catalysis A: General 2009. 353: p. 46-53.

[91] Bunch, A.Y. and U.S. Ozkan, Investigation of the Reaction Network of Benzofuran Hydrodeoxygenation over Sulfided and Reduced Ni–Mo/Al2O3 Catalysts. Journal of Catalysis 2002. 206: p. 177-187.

[92] Bunch, A.Y., X. Wang, and U.S. Ozkan, Hydrodeoxygenation of benzofuran over sulfided and reduced Ni–Mo/-Al2O3 catalysts: Effect of H2S. Journal of Molecular Catalysis A: Chemical 2007. 270: p. 264–272.

[93] Senol, O.İ., Hydrodeoxygenation of aliphatic andaromatic oxygenates on sulphided catalystsfor production of second generation biofuels, in Laboratory of Industrial Chemistry2007, Helsinki University of Technology: Finland.

Novel Biomass Utilization

Chapter 6

Sugarcane Biomass Production and Renewable Energy

Moses Isabirye, D.V.N Raju, M. Kitutu, V. Yemeline, J. Deckers and J. Poesen

Additional information is available at the end of the chapter

1. Introduction

Bio-fuel production is rooting in Uganda amidst problems of malnutrition and looming food insecurity "in [1,2]". The use of food for energy is a Worldwide concern as competition for resources between bio-fuel feedstocks and food crop production is inevitable. This is especially true for the category of primary feedstocks that double as food crops. Controversy surrounds the sustainability of bio-fuels as a source of energy in Uganda.

Given the above circumstances, adequate studies are required to determine the amount of feedstock or energy the agricultural sector can sustainably provide, the adequacy of land resources of Uganda to produce the quantity of biomass needed to meet demands for food, feed, and energy provision. Sugarcane is one of the major bio-fuel feed-stocks grown in Uganda.

Growth in sugarcane cultivation in Uganda is driven by the increased demand for sugar and related by-products. Annual sugar consumption in Uganda is estimated at 9 kg per capita with a predicted per capita annual consumption increase by 1 % over the next 15 years "in [3]".

This growth has resulted in increased demand for land to produce staple foods for households and thus encroaching on fragile ecosystems like wetlands, forests and shallow stoney hills and, a threat to food security " in [4]". The situation is likely to worsen with the advent of technology advancements in the conversion of biomass into various forms of energy like electricity and biofuels. A development that has attracted government and investors into the development of policies "in [5, 6]" that will support the promotion of bio-fuels in Uganda "in [7]".

Competition for land resources and conflicts in land use is imminent with the advent of developments in the use of agricultural crop resources as feedstocks for renewable energy production. Sugarcane is one such crop for which production is linked to various issues including the sustainability of households in relation to food availability, income and environmental integrity. The plans for government to diversify on altnertive sources of energy with focus on biofuels and electricity generation has aggravated the situation. This chapter aims at demonstrating how sugarcane biomass can sustainably be produced to support fuel and electrical energy demands while conserving the environment and ensuring increased household income and food security.

This study was conducted with a major objective of assessing sustainable production of bio-fuels and electricity from sugarcane biomass in the frame of household poverty alleviation, food security and environmental integrity.

2. Research methods

The assessment of sugarcane production potential is done for the whole country. The rest of the studies were done at the Sugar estate and the outgrower farmers.

2.1. Assessment of sugarcane production potential

The overall suitability assessment involved the use of the partial suitability maps of temperature, rainfall and soil productivity ratings (Figures 1 and 2). An overlay of the three maps gave suitability ratings for sugarcane bio-fuel feedstock.

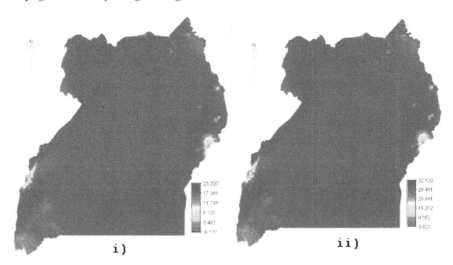

Figure 1. i) Minimum temperatures and maximum temperatures ii) " in [8] ".

Subtraction of gazetted areas, wetlands and water bodies produced final suitability maps and tables presented in the results. Steep areas have not been excluded since they are associated with highlands which are densely populated areas. It is hoped that soil conservation practices will be practiced where such areas are considered for production of sugarcane feedstock. Urban areas, though expanding, are negligible and have not been considered in the calculations.

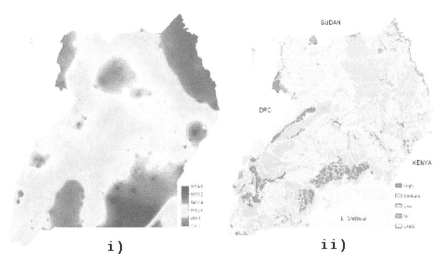

i)   **ii)**

Figure 2. i) Mean annual rainfall " in [8] "; ii) Soil productivity ratings " in [9] ".

The suitability of the land resource quality for sugarcane was based on sets of values which indicate how well each cane requirement is satisfied by each land quality say: mean annual rainfall, minimum and maximum temperatures and soil productivity. The four suitability classes (rating), assessed in terms of reduced yields, and were defined according to " in [10] ". Potential land-use conflict visualization also gives an indication of land available for the production of sugarcane bio-fuel feedstocks. Conflict visualization for food versus sugarcane was done by an overlay of suitability maps of maize with sugarcane. Land-use conflict with gazetted areas was assessed by overlaying gazetted area maps with sugarcane suitability map.

2.2. Sustaining sugarcane biomass productivity

The total biomass production of five commercial sugarcane varieties grown on the estate across all crop cycles (plant and three ratoons) was developed.

The data on cane yield and cane productivity of plant and ratoon crops between 1995-1996 and 2009-2010 were collected from annual reports and compared to experimental data. The total bio-mass (cane, trash and tops) production in plant and two ratoons were recorded at harvest age i.e. 18 and 17 months for plant and ratoon crops respectively. Data was collected from three locations of plot size 54 m² (4 rows of 10m length). Nutrient status of crop residues on oven dry basis was adopted as suggested by " see [11] " for calculation of nutrient return to the soil and nutrients available to succeeding crop.

A replicated trial with four replications was established during 2009-2010 with different levels of chemical fertilizers and factory by-products (filter mud and boiler ash) to test their influence on cane and sugar yields.

Semi-commercial trials were established on the estate to study the influence of green manuring with sunn hemp against no green manured blocks (control), aggressive tillage against reduced tillage, and intercropping with legumes on cane yield and juice quality parameters.

Field studies were conducted during 2002-2004 and 2007-2009 to evaluate the influence of different levels of Nitrogen (N), Phosphorus (P), Potassium (K) and sulphur (S) on cane yield and juice quality of plant crop of sugarcane.

The cane yield data on green cane vs burnt cane harvesting systems, and aggressive vs reduced tillage operations were collected and analysed for biomass yield.

2.3. Assessment of potential biofuel productivity and cane biomass electricity generation

Ethanol yield estimates from sugarcane is based on yield per ton of sugarcane. In addition, the production of bagasse from the cane stalk available for electricity generation were collected and analysed as per the following bagasse-steam-electrical power norms at Kakira sugar estate:

i.   Bagasse production is 40 % of sugarcane production
ii.   Moisture % in bagasse is 50%
iii. 1.0 ton of bagasse produces 2.0 tons of steam
iv.   5.0 tons of steam produces 1.0 Mwh electrical power

Therefore 2.5 tons of bagasse produces 5.0 tons of steam which will generate 1.0 Mwh electrical power. The electric power used in Kakira is hence generated from a renewable biomass energy source.

In 2005, Kakira had two 20 bar steam-driven turbo-generators (3 MW + 1.5 MW) in addition to 5 diesel standby generators. Thereafter, two new boilers of 50 tonnes per hour steam capacity at 45 bar-gauge pressure, with all necessary ancillaries such as an ash handling system, a feed water system and air pollution controls (such as wet scrubbers and a 40m 30 high chimney) were installed

2.4. Contribution to household income and food security

Indicative economic assessments included the use of gross sales for the raw material (farm gate) and ethanol. Annualized sugarcane net sales were compared to household annual expenditures to allow assessment of cane contribution to household income. Integration of commodity prices gives insight on the potential contribution of bio-fuels to household poverty alleviation and overall development of rural areas.

3. Results

3.1. Suitability

The agro-ecological settings favor the growing of sugarcane with a potential 10,212,757 ha (49.6%) at a marginal level of production with 2,558,698 ha (12.4%) land area potentially not suitable for cane production. Although the current production is far below the potential production "in [12]", the related cane production is 908,935,330 and 60,769,069 tons respectively. It is also evident that there is possibility of increasing production through expansion of land area under sugarcane.

1. Suitability of sugarcane production and conflict visualization between food crops and gazetted areas in Uganda

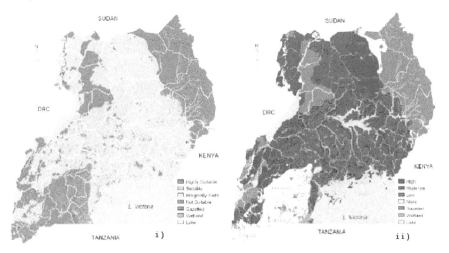

Figure 3. Sugar cane suitability ratings (i) and conflict visualization between food crops and gazetted areas

The marginal productivity of cane in Uganda is a function of Rainfall amount and the atmospheric temperature. Nevertheless the average optimum yields (89 ton / ha) at marginal level of productivity are comparable to yields of 85 ton / ha in a commercialized production in Brazil "in [13] ".

Expanding acrage under sugarcane is likely to increase pressure on gazetted biodiversity rich areas including wetlands with consequent potential loss of bio-diversity.

Sugarcane and maize (food crop) have similar ecological requirements, presenting a situation of high potential land-use conflict as 49.6 % of arable land can be grown with both sugarcane and food crops (figure 3 i). Figure 3 ii), shows 14 % of the land where sugarcane has potential conflict with gazetted areas of which 4.3 % has potential conflict with forest reserves.

Sugarcane, given its energy balance advantage, is likely to be beneficial if promoted as bio-fuel feedstock as this is likely to increase sugarcane prices to the benefit of the small scale farmer.

3.2. Agronomy

The beneficial effects of integrated agronomic practices like reduced tillage operations, balanced fertilization; organic recycling of mill by-products (filter mud and boiler ash); intercropping with legumes; green manuring with sunn hemp; crop residue recycling through cane trash blanketing in ratoons by green cane harvesting to sustain soil fertility and cane productivity in monoculture sugarcane based cropping system are presented and discussed. Partitioning of dry matter between plant and ratoon crops of cane grown on the estate and Outgrowers fields were quantified and also presented in this chapter.

3.2.1. Influence of agronomic practices on cane yield and cane productivity

3.2.1.1. Green manuring

There was considerable increase in cane yield (7.92 tc/ha) and cane productivity (0.62 tc/ha/m) in plant and ratoon crops due to green manuring as compared to blocks without green manuring (Table 1).

Crop cycle	Green manuring		No green manuring		Variance	
	Yield tc/ha	Productivity tc/ha/m	Yield tc/ha	Productivity tc/ha/m	Yield tc/ha	Productivity tc/ha/m
Plant	125.3	5.9	112.2	5.2	13.1	0.8
Ratoon 1	95.3	5.8	92.4	5.0	2.8	0.8
Ratoon 2	92.4	5.0	84.6	4.7	7.8	0.3
# average	104.3	5.6	96.4	5.0	7.9	0.6

Table 1. Cane yield and cane productivity variance due to green manuring

Growing sunn hemp (Crotolaria juncia) during fallow period for in-situ cultivation has been a common practice to improve soil health on the estate since 2004. Sunn hemp at 50% flowering on average produces 27.4 t/ha and 5.9 t/ha of fresh and dry weights respectively. It contains 2.5% N on oven dry basis and adds about 147kg N/ha to the soil. Of this amount, 30% (44kg N/ha) is presumed to be available to the succeeding sugarcane plant crop. "[11]" reported that N available to sugarcane ranges between 30 - 60% of total N added to soils in South Africa.

3.2.1.2. Balanced fertilization

The results indicated that application of N to plant crop at 100kg /ha, phosphorus at 160kg P_2O_5 /ha, potassium at 100 K_2O /ha and sulphur at 40kg /ha significantly increased the cane yields by 23.3 tc/ha; 22.25 tc/ha; 12.07 tc/ha and 8.71 tc/ha respectively over no application of

N, P, K and S. Sugar yields were also improved due to N, P, K and S application by 2.91 ts/ha; 2.17 ts/ha; 2.88 ts/ha and 0.44 ts/ha respectively as compared to no application (Table 2).

Nutrient levels (kg/ha)	Attributes	
	Cane yield (tc/ha)	Sugar yield (ts/ha)
N levels: 0	112.68	15.69
50	124.38	17.08
100	135.98	18.60
P_2O_5 levels: 0	108.70	14.51
80	120.40	16.34
160	130.95	18.01
K_2O levels: 0	108.06	15.58
50	124.86	17.32
100	134.13	18.46
Sulphur levels: 0	113.90	11.92
40	122.61	12.36
CD at 5%: N	10.69	1.20
P_2O_5	9.80	1.18
K_2O	9.06	1.06
S	6.94	0.32

Table 2. Influence of N, P, K and S nutrition on cane and sugar yields

Balanced fertilizer application is very vital for crop growth. Adequate amounts of especially the major nutrients need to be supplied for proper crop growth. Excessive application of N in cane plant crop has been shown to inhibit the activity of free living N-fixing bacteria and chloride ions from Muriate of potash adversely affecting soil microbial populations "in [14]".

3.2.1.3. Mill by-products

Millable stalk population at harvest was significantly higher due to application of filter mud and boiler ash + 100% recommended dose of fertilizers (RDF) than all other treatment combinations. However, there was no significant effect on stalk length, number of internodes and stalk weight due to different treatments. Cane and sugar yields were significantly higher by 30.2 tc/ha and 3.8 ts/ha respectively due to application of filter mud and boiler ash + 100% RDF. The data are presented in Table 3.

Cane-mill by-products (filter mud and boiler ash) do contain valuable amounts of N, P, K, Ca, and several micronutrients "in [15]". These in addition to the inorganic fertilizers applied considerably increased cane yields as compared to treatments which received only inorganic fertilizers. In [11] it is also showed that organic wastes-including filter mud and bolier ash could be used as an alternative source of nutrients in cane cultivation.

Treatment	Yield (T/ha)	
	Cane	Sugar (Estimated)
1. 100% recommended dose of fertilizers (RDF)	131.6	16.6
2. Filter mud + Boiler ash @ 32 + 8 T/ha alone	122.1	16.2
3. 100% RDF+ Filter mud + Boiler ash @ 32 + 8 T/ha	161.7	20.4
4. 75% RDF+ Filter mud + Boiler ash @ 32 + 8 T/ha	143.8	17.5
5. 50% RDF+ Filter mud + Boiler ash @ 32 + 8 T/ha	142.1	17.3
6. 25% RDF+ Filter mud + Boiler ash @ 32 + 8 T/ha	135.0	16.6
CD @ 5%	12.3	1.9

Table 3. Influence of mill by-products on cane and sugar yields

3.2.1.4. Partitioning of bio-mass

Among the crop cycles, plant crop recorded higher bio-mass production than succeeding ratoons. The production of crop residues were also higher (50.7 t/ha) in plant crop than 1st (45.7 t/ha and 2nd (36.6 t/ha) ratoons. The data are presented in the Table 4 below.

Crop cycle	Cane weight (Tc/ha)	Tops weight (T/ha)	Trash weight (T/ha)	Total bio-mass (T/ha)	Tops + Trash weight T/ha	% over total
Plant	137.5	30.3	20.4	188.2	50.7	27.0
Ratoon 1	124.9	26.8	18.9	170.59	45.7	27.0
Ratoon 2	107.5	19.7	16.8	144.02	36.6	25.4
# average	123.3	Fresh: 25.6 Dry: 9.0	Fresh: 18.7 Dry: 16.9	- -	Fresh: 44.3 Dry: 25.8	26.4

Table 4. Crop-wise partitioning of bio-mass

Results indicate that on average, 25.8 tons of dry matter (cane trash and tops) is produced from each crop cycle at harvest. In burnt cane harvesting system, all this dry matter is lost unlike in green cane harvesting. This explains the gradual decline in cane yield in such harvesting systems. The decline is presumed to be due to deteriorating organic matter and other physical and chemical properties of the soil " in [11] " .

After decomposition of cane trash, 139kg N, 59kg P_2O_5, 745kg K_2O, 41kg Ca, 46kg Mg and 34kg S /ha were added to the soils and these added nutrients would be available to the succeeding crop at 30% of the total nutrients " in [11] " . The nutrient concentration of crop residues (trash + tops) were taken into account for computing nutrient additions to the soil and their availability to the succeeding ratoon crops and the data are presented in Table 5.

Nutrient status of crop residues (%)	Total dry matter (T/ha)	Total nutrients	
		Added to soil (kg/ha)	Available to crops @ 30% (kg/ha)
N : 0.54	25.8	139	42
P₂O₅ : 0.23	25.8	59	18
K₂O : 2.89	25.8	745	223
Ca : 0.16	25.8	41	12
Mg : 0.18	25.8	46	14
S : 0.13	25.8	34	10

** Adopted from [11]

Table 5. Nutrient status of crop residues and their availability to the succeeding crops

3.3. Renewable energy potential

3.3.1. Ethanol productivity

Sugarcane, given its energy balance advantage, is likely to be beneficial if promoted as bio-fuel feedstock as this is likely to increase sugarcane prices to the benefit of the small scale farmer.

Promoting sugarcane as a feedstock for ethanol is likely to improve rural livelihood and also minimize on forest encroachment since energy output per unit land area is very high for sugarcane.

In Brazil for example the production of sugar cane for ethanol only uses 1% of the available land and the recent increase in sugar cane production for bio-fuels is not large enough to explain the displacement of small farmers or soy production into deforested zones " in [13] ".

To minimize competition over land, it is advisable to grow sugarcane that has high yields with higher energy output compared to other biofuel crops. High yielding bio-fuels are preferable as they are less likely to compete over land "in [16]".

3.3.2. Bioelectricity generation

Hydropower contributes about 90 per cent of electricity generated in Uganda with sugarcane based bagasse bioelectricity, fossil fuel and solar energy among other sources of power. Although the current generation of 800 MW "in [18] ". has boosted industrial growth, the capacity is still lagging behind the demand that is driven by the robust growth of the economy.

The low pressure boilers of 45 bar currently generate 22 MW of which 10 MW is connected to the grid. However, Kakira sugar estate has a target of generating 50 MW of electricity with the installation of higher pressure boilers of 68 bar in 2013. This target can be surpassed given the abundance of the bagasse (Table 6 and 7).

Particulars	Plant	Ratoon 1	Ratoon 2	Ratoon 3	Total/average
Cane harvest area (ha)	1,461.1	1,541.7	1,535.2	392.8	4,930.8
Total cane supply (tons)	155,207.3	162,132.3	137,436.4	40,138.9	494,915.9
Average cane yield (tc/ha)	106.23	105.16	89.52	87.28	100.37
Average harvest age (months)	19.20	18.15	17.98	16.50	17.96
Cane productivity (tc/ha/m)	5.53	5.79	4.97	5.29	5.40
Bagasse production /ha	42.49	42.06	35.80	34.90	40.14
Steam generation (tons /ha)	84.98	84.12	71.60	69.80	80.28
Electric power generation (Mwh/ha)	17.00	16.82	14.32	13.96	16.05

*This electric power generation is calculated based on using low pressure boilers of 45 bar at Kakira estate

Table 6. Mean estate cane production/productivity, electrical generation* norms (2008 – 2012)

Particulars	Plant	Ratoon 1	Ratoon 2	Ratoon 3	Total/average
Cane harvest area (ha)	3,429.3	3,206.9	2,334.7	1,431.3	10,402.2
Total cane supply (tons)	320,100.20	290,073.29	188,274.65	113,837.25	912,285.49
Average cane yield (tc/ha)	93.34	90.45	80.64	79.53	87.70
Average harvest age (months)	18.50	17.50	18.00	16.00	17.50
Cane productivity (tc/ha/m)	5.05	5.17	4.48	4.97	5.01
Bagasse production /ha	37.30	36.18	32.25	31.81	35.08
Steam generation (tons /ha)	74.60	72.36	64.50	63.62	70.16
Electric power generation (Mwh/ha)	14.90	14.50	12.90.	12.70	14.00

*This electric power generation is calculated based on using low pressure boilers of 45 bar at Kakira estate

Tc = Tons of cane

Table 7. Mean outgrowers cane production/productivity, electrical generation* norms (mean for 2008 – 2012)

Putting into consideration the productivity norms at Kakira estate and outgrowers (Table 8), with a potential of producing 908.9 m tons of sugarcane, Uganda has a potential of producing bio-electricity that surpasses the nation's demand by far. Much of this electrical power can be exported to the region, greatly expanding on Uganda's export base.

Particulars	Plant	Ratoon 1	Ratoon 2	Ratoon 3	Total/average
Cane harvest area (ha)	4,890.4	4,748.6	3,869.9	1,824.1	15,333.0
Total cane supply (tons)	475,307.50	452,205.60	325,711.05	153,976.15	1,407,200.4
Average cane yield (tc/ha)	97.19	95.22	84.17	84.41	91.78
Average harvest age (months)	18.75	17.75	18.00	16.00	17.65
Cane productivity (tc/ha/m)	5.18	5.36	4.67	5.27	5.20
Bagasse production /ha	38.88	38.09	33.67	33.76	36.71
Steam generation (tons /ha)	77.76	76.18	67.34	67.52	73.42
Electric power generation (Mwh/ha)	15.55	15.24	13.47	13.50	14.68

*This electric power generation is calculated based on using low pressure boilers of 45 bar at Kakira estate

Table 8. Combined (Estate + Outgrowers cane production/productivity, Electrical power generation* norms (mean for 2008 – 2012)

3.4. Household income and food security

The competition for resources between sugarcane and food crops is apparent with foreseen consequent increased food insecurity. Fifty percent of the arable land area good for food crop production is equally good for sugarcane.

A farm household that allocates all of its one hectare of land to sugarcane is expected to earn 359 $ at high input level, 338 $ at intermediate level and 261 $ at low input level " in [4] ". The 391 $ required to purchase maize meal is well above the net margins from one ha. This shows that proceeds from one hectare cannot sustain a household of 5. It is further revealed that maize produced from 0.63 ha can sustain a household nutritionally; however considering the annual household expenditure (760.8 $; "in [17] "), about three hectares of land under sugarcane are required at low input level to support a household " in [4]".

However, this study reveals that sugarcane sales accrued from ethanol under a scenario of a flourishing bio-fuel industry is associated with increased income that is likely to support households (Table 9). An ethanol gross sale per person per day is 1.6 dollars; an indication that the cultivation of sugarcane based biofuel is likely to contribute to alleviation of household poverty. A trickle-down effect on household income is expected from a foreseen expansion of bagasse-based electricity generation beyond the estate into the national electricity grid.

Production / year		Gross sales			Conflict		
Cane production	Billion	Farm	Ethanol	Capita	Food	Gazetted	Forest
ton	litres	/ha/year			%		
908.9 m	75.4	1869	22161	1.6	50.0	14.0	4.3
Sugarcane= USD 21/ton: projected population of 33 m in 2009 is used							

Table 9. Sugarcane productivity, sales and potential land-use conflict

4. Further research

The expansion of cane production is largely driven by market forces oblivious to the detrimental impact the industry is likely to have on food, livelihood security and the status of biodiversity. In addition to lack of appropriate policies to support the small-scale cane farmer, the policies are largely sectoral with no linkages with other relevant policies. Information is required to support the sustainable development of the cane industry with minimal negative impact on food and livelihood security and the status of biodiversity.

5. Relevant questions to explore among others include

Can food crop productivity be improved in the context of a sugarcane-based farming system?

Can the understanding of the dimensions of food and livelihood security in sugarcane-based farming systems inform the synergistic development and review of relevant policies in the food, agriculture, health, energy, trade and environment sectors? What are the social impacts of the industry in light of the various agro-ecological zones of the country? What is the gender based livelihood strategies with special emphasis on labor exploitations- child labor etc?

What do people consider as possible options for improving food and livelihood security in a sugarcane-based farming system? Do these options differ between different actors (local women and men, NGOs and government)? How do families cope with food inadequacy, inaccessibility and malnutrition?

Can the study inform the carbon credit market initiative for farming systems in Uganda through the climate smart agriculture concept? Are the proposed assessment tools appropriate for Ugandan situations and the cane-based systems in particular?

6. Conclusion

Driven by the need to meet the increasing local and regional sugar demand, and fossil fuel import substitution, cane expansion has potential negative impact on food security and biodiversity. However, this negative impact parallels the benefits related to cane cultivation. Cane biomass yield can be improved and sustained through the integrated use of various practices reported in this study. Consequently this reduces the need to expand land acreage under cane while releasing land for use in food crop productivity. The high biomass returned to the ground sequesters carbon thereby offering the opportunity for sugarcane based farmers to earn extra income through the sale of carbon credits. Trickle down effects are expected to increase household income through the production and marketing of cane based biofuel and electricity.

These developments are expected to improve the farmers purchasing power, making households to be less dependent on the land and more food secure financially.

Author details

Moses Isabirye*
Faculty of Natural Resources and Environment, Namasagali Campus, Busitema University, Kamuli, Uganda

D.V.N Raju
Research and Dev't Section - Agricultural Department Kakira Sugar Limited, Jinja, Uganda

M. Kitutu
National Environment Management Authority, Kampala, Uganda

V. Yemeline
UNEP/GRID-Arendal, Norway

J. Deckers and J. Poesen
Katholieke Universiteit Leuven, Department of Earth and Environmental Sciences, Celestijnenlaan Leuven, Belgium

Acknowledgement

Financial support for the various studies reported in this chapter was provided by UNEP GRID-Arendal, Kakira Sugar Limited and Belgian Technical Cooperation (BTC). The National Environment Management Authority and the National Agricultural Research Organization in Uganda gave the facilities and technical support that enabled the accomplishment of studies reported here.

7. References

[1] Bahiigwa B. A. Godfrey, 1999 Household Food Security In Uganda: An Empirical Analysis, Economic Policy Research Center, Kampala, Uganda
[2] Uganda government, 2002b Uganda Food and Nutrition Policy, Ministry of Agriculture, Animal Industry and Fisheries, Ministry of Health, Kampala, Uganda

* Corresponding Author

[3] USCTA (2001) The Uganda Sugarcane Technologist's Association, Fourth Annual Report, 2001, Kakira, Uganda.

[4] Isabirye (2005) Land Evaluation around Lake Victoria: Environmental Implications for Land use Change, PhD Dissertation, Katholieke Universiteit, Leuven, Belgium

[5] MEMD (2007) The Renewable Energy Policy for Uganda. MEMD, Kampala, Uganda.

[6] MEMD (2010) The Uganda Energy Balance Report., MEMD, Kampala, Uganda.

[7] Bio-fuel-news (2009) Uganda to produce cellulosic ethanol in a year. Bio-fuel International volume 3, issue 10,2009, http://www.Bio-fuel-news.com/magazine_store.php?issue_id=34

[8] Meteorology Department (1961) Climate data, Meteorological Department, Entebbe Uganda

[9] Chenery (1960) Introduction to the soils of the Uganda Protectorate, Memoirs of the Research Division, Series 1- Soils, Number 1, Department of Agriculture, Kawanda Research Station, Uganda

[10] FAO (1983) Guidelines: Land evaluation for rainfed agriculture. FAO Soils Bulletin 52, Food and Agricultural Organization of the United Nations, Rome

[11] Antwerpen R. V. (2008) Organic wastes as an alternative source of nutrients. The link published by SASRI. Vol. 17 No. 2: May 2008. 8-9.

[12] FAOStat (2012) http://faostat.fao.org/site/339/default.aspx Sunday, May 06, 2012

[13] Xavier M.R. (2007) The Brazilian sugarcane ethanol experience. Issue Analysis, no. 3, Washington, USA, Competitive Enterprise Institute. 11 p.

[14] Carr Carr, A.P, Carr, D.R, Carr, I.E, Wood, A.W and Poggio, M. (2008) Implementing sustainable farming practices in the Herbert: The Oakleigh farming company experience

[15] Raju, D.V.N and Raju, K.G.K (2005) Sustainable sugarcane production through integrated nutrient management. In: Uganda Sugarcane Technologists' Association 17TH Annual Technical Conference.

[16] Pesket Leo, Rachel Slater, Chris Steven, and Annie Dufey (2007) Biofuels, Agriculture and Poverty Reduction. Natural Resource Perspectives 107, Overseas Development Institute, 111 Westminster Bridge Road, London SE1 7JD

[17] UBOS (2001) Uganda National household survey 1999/2000; Report on the socioeconomic. Uganda Bureau of Statistics, Entebbe, Uganda. www.ubos.org

[18] Ibrahim Kasita (2012) Strategic plan to increase power supply pays dividends, New Vision, Publish Date: Oct 09, 2012

Towards the Production of Second Generation Ethanol from Sugarcane Bagasse in Brazil

T.P. Basso, T.O. Basso, C.R. Gallo and L.C. Basso

Additional information is available at the end of the chapter

1. Introduction

Brazil and the United States produce ethanol mainly from sugarcane and starch from corn and other grains, respectively, but neither resource are sufficient to make a major impact on world petroleum usage. The so-called first generation (1G) biofuel industry appears unsustainable in view of the potential stress that their production places on food commodities. On the other hand, second generation (2G) biofuels produced from cheaper and abundant plant biomass residues, has been viewed as one plausible solution to this problem [1]. Cellulose and hemicellulose fractions from lignocellulosic residues make up more than two-thirds of the typical biomass composition and their conversion into ethanol (or other chemicals) by an economical, environmental and feasible fermentation process would be possible due to the increasing power of modern biotechnology and (bio)-process engineering [2].

Brazil is the major sugar cane producer worldwide (ca. 600 million ton per year). After sugarcane milling for sucrose extraction, a lignocellulosic residue (sugarcane bagasse) is available at a proportion of ca. 125 kg of dried bagasse per ton of processed sugarcane. Therefore, sugarcane bagasse is a suitable feedstock for second generation ethanol coupled to the first generation plants already in operation, minimizing logistic and energetic costs.

State-owned energy group Petrobras is one of the Brazilian groups leading the development of second generation technologies, estimating that commercial production could begin by 2015. Other organizations making significant contributions to next generation biofuels in Brazil include the Brazilian Sugarcane Technology Centre (CTC), operating a pilot plant for the production of ethanol from bagasse in Piracicaba (Sao Paulo) [3]. Recently, GraalBio (Grupo Brasileiro Graal) has stated publically that will start production of bioethanol from sugarcane bagasse in one plant located at the Northeast region of the country.

Pretreatment, hydrolysis and fermentation can be performed by various approaches. According to a CTC protocol, the process of manufacturing ethanol from bagasse is divided into the following steps. First, the bagasse is pretreated via steam explosion (with or without a mild acid condition) to increase the enzyme accessibility to the cellulose and promoting the hemicellulose hydrolysis with a pentose stream. The lignin and cellulose solid fraction is subjected to cellulose hydrolysis, generating a hexose-rich stream (mainly composed of glucose, manose and galactose). The final solid residue (lignin and the remaining recalcitrant cellulose) is used for heating and steam generation. The hexose fraction is mixed with 1G cane molasses (as a source of minerals, vitamins and aminoacids) and fermented by regular *Saccharomyces cerevisiae* industrial strains (not genetically modified) using the same fermentation and distillation facilities of the Brazilian ethanol plants. The pentose fraction will be used as substrate for other biotechnological purposes, including ethanol fermentation.

Researchers are focusing on cutting the cost of the enzymes and the pretreatment process, as well as reducing energy input. Production of ethanol from sugarcane bagasse will have to compete with the use of bagasse for electricity cogeneration. Depending on the efficiency of the cogeneration plant, about half of the bagasse is required to produce captive energy in the form of steam and energy at the sugar and ethanol facility. It is estimated that the surplus bagasse could increase the Brazilian ethanol production by roughly 50% [3].

2. Lignocellulosic residues

Lignocellulosic residues are composed by three main components: cellulose, hemicellulose and lignin. Cellulose and hemicellulose are polyssacharydes composed by units of sugar molecules [4]. Sugarcane bagasse is composed of around 50% cellulose, 25% hemicellulose and 25% lignin. It has been proposed that because of its low ash content (around 2-3%), this product offers numerous advantages in comparison to other crop residues (such as rice straw and wheat straw) when used for bio-processessing purposes. Additionally, in comparison to other agricultural residues, bagasse is considered a richer solar energy reservoir due to its higher yields on mass/area of cultivation (about 80 t/ha in comparison to about 1, 2, and 20 t/ha for wheat, other grasses and trees, respectively) [5].

2.1. Cellulose

Cellulose is composed of microfibrils formed by glucose molecules linked by β-1,4, being each glucose molecule reversed in relation to each other. The union of microfibrils form a linear and semicrystalline structure. The linearity of the structure enables a strong bond between the microfibrils. The crystallinity confers resistance to hydrolysis due to absence of water in the structure and the strong bond between the glucose chains prevents hydrolases act on the links β -1,4 [1].

2.2. Hemicellulose

Hemicellulose is a polysaccharide made of polymers formed by units of xylose, arabinose, galactose, manose and other sugars, that present crosslinking with glycans. Hemicellulose can bind to cellulose microfibrils by hydrogen bonds, forming a protection that prevents the contact between microfibrils to each other and yielding a cohesive network. Xyloglucan is the major hemicellulose in many primary cell walls. Nevertheless, in secondary cell wall, which predominate in the plant biomass, the hemicelluloses are typically more xylans and arabino-xylans. Typically, hemicellulose comprises between 20 to 50% of the lignocellulose polysaccharides, and therefore contributes significantly to the production of liquid biofuels [1]. Sugarcane bagasse contain approximately 25-30% hemicelluloses [13,14].

2.3. Lignin

Lignin is a phenolic polymer made of phenylpropanoid units [13], which has the function to seal the secondary cell wall of plants. Besides providing waterproofing and mechanical reinforcement to the cell wall, lignin forms a formidable barrier to microbial digestion. Lignin is undoubtedly the most important feature underlying plant biomass architecture. Sugarcane bagasse and leaves contain approximately 18-20% lignin [13]. The phenolic structure of this polymer confers a material that is highly resistant to enzymatic digestion. Its disruption represents the main target of pretreatments before enzymatic hydrolysis.

3. Pretreatment

The pretreatment process is performed in order to separate the carbohydrate fraction of bagasse and other residues from the lignin matrix. Another function is to minimize chemical destruction of the monomeric sugars [6]. During pretreatment the inner surface area of substrate particles is enlarged by partial solubilization of hemicellulose and lignin.. This is achieved by various physical and/or chemical methods [5]. However, it has been generally accepted that acid pretreatment is the method of choice in several model processes [7]. One of the most cost-effective pretreatments is the use of diluted acid (usually between 0.5 and 3% sulphuric acid) at moderate temperatures. Albeit lignin is not removed by this process, its disruption renders a significant increase in sugar yield when compared to other processes [1]. Regarding sugarcane bagasse several attempts have been made to optimize the release of the carbohydrate fraction from the lignin matrix, including dilute acid pretreatment, steam explosion, liquid hot water, alkali, peracetic acid and also the so called ammonia fiber expansion (AFEX).

4. Hydrolysis

Cellulose and hemicellulose fractions released from pretreatment has to be converted into glucose and other monomeric sugars. This can be achieved by both chemically- or enzymatically- oriented hydrolysis [7, 8].

4.1. Chemical hydrolysis

Whitin chemical hydrolysis, acid hydrolysis is the most used and it can be performed with several types of acids, including sulphurous, sulphuric, hydrochloric, hydrofluoric, phosphoric, nitric and formic acid. While processes involving concentrated acids are usually operated at low temperatures, the large amount of acids required may result in problems associated with equipment corrosion and energy demand for acid recovery. These processes typically involve the use of 60-90% concentrated sulfuric acid. The primary advantage of the concentrated acid process in realtion to diluted acid hydrolysis is the high sugar recovery efficiency, which can be on the order of 90% for both xylose and glucose. Concentrated acid hydrolysis disrupts the hydrogen bonds between cellulose chains, converting it into a completely amorphous state [8].

On the other hand, during dilute acid hydrolysis temperatures of 200–240°C at 1.5% acid concentrations are required to hydrolyze the crystalline cellulose. Besides that, pressures of 15 psi to 75 psi, and reaction time in the range of 30 min to 2 h are employed. During this conditions degradation of monomeric sugars into toxic compounds and other non-desired products are inevitable [9]. The main advantage of dilute acid hydrolysis in comparison to concentrated acid hydrolysis is the relatively low acid consumption. However, high temperatures required to achieve acceptable rates of conversion of cellulose to glucose results in equipment corrosion [4].

4.2. Enzymatic hydrolysis

Differently from acid hydrolysis, biodegradation of sugarcane bagasse by cellulolytic enzymes can be performed at much lower temperatures (around 50°C or even lower). Moreover conversion of cellulose and hemicellulose polymers into their constituent sugars is very specific and toxic degradation products are unlikely to be formed. However, a pretreatment step is required for enzymatic hydrolysis, since the native cellulose structure is well protected by the matrix compound of hemicellulose and lignin [4].

Cellulase is the general term for the enzymatic complex able to degrade cellulose into glucose molecules. The mechanism action accepted for hydrolysis of cellulose are based on synergistic activity between endoglucanase (EC 3.2.1.4), exoglucanase (or cellobiohydrolase (EC 3.2.1.91)), and β-glucosidase (EC 3.2.1.21). The first enzyme cleaves the bounds β-1,4-glucosidic of cellulose chains to produce shorter cello-dextrins. Exoglucanase release cellobiose or glucose from cellulose and cello-dextrin chains and, finally β-glucosidases hydrolyze cellobiose to glucose. The intramolecular β-1,4-glucosidic linkages are cleaved by endoglucanases randomly. Endoglucanases and exoglucanases have different modes of action. While endoglucanase hydrolyze intramolecular cleavages, exoglucanases hydrolyze long chains from the ends. More specifically, exoglucanases or cellobiohydrolases have action on the reducing (CBH I) and non-reducing (CBH II) cellulose chain ends to liberate glucose and cellobiose. These enzymes acts on insoluble cellulose, then their activity are often measured using microcrystalline cellulose. Lastly, β-glucosidases or cellobioase hydrolyze cellobiose to glucose. They are important to the process of hydrolysis because they removed cellobiose to the aqueous phase that is an inhibitor to the action of endoglucanases and exoglucanases [10].

The multi-complex enzymatic cocktail known as cellulase and hemicellulase can be produced by a variety of saprophytic microorganisms. *Trichoderma* and *Aspergillus* are the genera most used to produce cellulases. Among them, one of the most productive of biomass degrading enzymes is the filamentous fungus *Trichoderma reesei*. It cellulolytic arsenal is composed by a mixture of endoglucanases and exoglucanases that act synergistically to break down cellulose to cellobiose. Two β-glucosidases have been identified that are implicated in hydrolyzing cellobiose to glucose. An additional protein, swollenin, has been described that disrupts crystalline cellulose structures, presumably making polysaccharides more accessible to hydrolysis. The four most abundant components of *T. reesei* cellulase together constitute more than 50% of the protein produced by the cell under inducing conditions [9, 15]. Cellulases are essential for the biorefinery concept. In order to reduce the costs and increase production of commercial enzymes, the use of cheaper raw materials as substrate for enzyme production and focus on a product with a high stability and specific activity are mandatory. Apart from bioethanol, there are several applications to these enzymes, such as in textile, detergent, food, and in the pulp and paper industries.

5. Enzyme production using sugarcane bagasse

Cultivation of microorganisms in agroindustrial residues (such as bagasse) aiming the production of enzymes can be divided into two types: processes based on liquid fermentation or submerged fermentation (SmF), and processes based on solid-state fermentation (SSF) [5]. In several SSF processes bagasse has been used as the solid substrate. In the majority of the processes bagasse has been used as the carbon (energy) source, but in some others it has been used as the solid inert support. Cellulases have been extensively studied in SSF using sugarcane bagasse. It has been reported the production of cellulases from different fungal strains [5].

Several processes have been reported for the production of enzymes using bagasse in SmF. One of the most widely studied aspects of bagasse application has been on cellulolytic enzyme production. Generally basidiomycetes have been employed for this purpose, in view of their high extracellular cellulase production. A recent example was the use of Trichoderma reesei QM-9414 for cellulase and biomass production from bagasse. Additionally, white-rot fungi were successfully used for the degradation of long-fiber bagasse. Most of the strains caused an increase in the relative concentration of residual cellulose, indicating that hemicellulose was the preferred carbon source [5].

6. Processes for ethanol production

Ethanol production from lignocellulosic residues is performed by fermentation of a mixture of sugars in the presence of inhibiting compounds, such as low molecular weight organic acids, furan derivatives, phenolics, and inorganic compounds released and formed during pretreatment and/or hydrolysis of the raw material. Ethanol fermentation of pentose sugars (xylose, arabinose) constitutes a challenge for efficient ethanol production from these residues, because only a limited number of bacteria, yeasts, and fungi can convert pentose (xylose, arabinose), as wells as other monomers released from hemicelluloses (mannose, galactose) into ethanol with a satisfactory yield and productivity [8]. Hydrolysis and fermentation processes can be designed in various configurations, being performed separately, known as separate hydrolysis and fermentation (SHF) or simultaneously, known as simultaneous saccharification and fermentation (SSF) processes.

When hydrolysis of pretreated cellulosic biomass is performed with enzymes, these biocatalysts (endoglucanase, exoglucanase, and β-glucosidase) can be strongly inhibited by hydrolysis products, such as glucose, xylose, cellobiose, and other oligosaccharides. Therefore, SSF plays an important role to circumvent enzyme inhibition by accumulation of these sugars. Moreover, because accumulation of ethanol in the fermenters does not inhibit cellulases as much as high concentrations of sugars, SSF stands out as an important strategy for increasing the overall rate of cellulose to ethanol conversion. Some inhibitors present in the liquid fraction of the pretreated lignocellulosic biomass also have a significant and negative impact on enzymatic hydrolysis. Due to the decrease in sugar inhibition during enzymatic hydrolysis in SSF, the detoxifying effect of fermentation, and the positive effect of some inhibitors present in the pretreatment hydrolysate (e.g. acetic acid) on the fermentation, SSF can be an advantageous process when compared to SHF [8]. Another important advantage is a reduction in the sensitivity to infection in SSF when compared to SHF. However, this was demonstrated by Stenberg and co-authors [12] that this is not always the case. It was observed that SSF was more sensitive to infections than SHF. A major disadvantage of SSF is the difficulty in recycling and reusing the yeast since it will be mixed with the lignin residue and recalcitrant cellulose.

In SHF, hydrolysis of cellulosic biomass and the fermentation step are carried out in different units [8], and the solid fraction of pretreated lignocellulosic material undergoes hydrolysis (saccharification) in a separate tank by addition of acids/alkali or enzymes. Once hydrolysis is completed, the resulting cellulose hydrolysate is fermented and the sugars converted to ethanol. *S. cerevisiae* is the most employed microorganism for fermenting the hydrolysates of lignocellulosic biomass. This yeast ferments the hexoses contained in the hydrolysate but not the pentoses. Several strategies (screening biodiversity, metabolic and evolutionary engineering of microorganisms) have been attempted to overcome this metabolic limitation.

One of the main features of SHF process is that each step can be performed at its optimal operating conditions (especially temperature and pH) as opposed to SSF [7].Therefore, in SHF each step can be carried out under optimal conditions, i.e. enzymatic hydrolysis at 45-50°C and fermentation at 30-32°C. Additionally, it is possible to run fermentation in continuous mode with cell recycling. The major drawback of SHF, as mentioned before, is that the sugars released during hydrolysis might inhibit the enzymes. It must be stressed out that ethanol produced can also act as an inhibitor in hydrolysis but not as strongly as the sugars. A second advantage of SSF over SHF is the process integration obtained when hydrolysis and fermentation are performed in one single reactor, which reduces the number of reactors needed.

Experimental data from ethanol output using sugarcane bagasse as the substrate is being released in the literature. Vásquez et al. (2007) [13] described a process in which 30 g/L of ethanol was produced in 10h fermentation by *S. cerevisiae* (baker's yeast) at an initial cell concentration of 4 g/L (dry weight basis) using non-supplemented bagasse hydrolysate at 37°C. The hydrolysate was obtained by a combination of acid/alkali pre-treatment, followed by enzymatic hydrolysis. Krishnan et al. (2010) [14] reported 34 and 36 g/L of ethanol on AFEX-treated bagasse and cane leaf residue, respectively, using recombinant *S. cerevisiae* and 6% (w/w) glucan loading during enzymatic hydrolysis. Overall the whole process produced around 20kg of ethanol per 100kg of each bagasse or cane leaf, and was performed during 250h including pretreatment, hydrolysis and fermentation. According to the authors this is the first complete mass balance on bagasse and cane leaf.

In view of the growing concern over climate change and energy supply, biofuels have received positive support from the public opinion. However, growing concern over first generation biofuels in terms of their impact on food prices and land usage has led to an increasing bad reputation towards biofuels lately. The struggle of 'land vs fuel' will be driven by the predicted 10 times increase in biofuels until 2050. The result is that biofuels are starting to generate resistance, particularly in poor countries, and from a number of activist non-governmental organizations with environmental agendas. This is highly unfortunate as it is clear that liquid biofuels hold the potential to provide a more sustainable source of energy for the transportation sector, if produced sensibly. Since replacement of fossil fuels will take place soon, a way to avoid these negative effects from first generation biofuels (mainly produced from potential food sources) is to make lignocellulosic derived fuels available within the shortest possible time. It is known that this process involves an unprecedented challenge, as the technology to produce these replacement fuels is still being developed [1]. Fuels derived from cellulosic biomass are essential in order to overcome our excessive dependence on petroleum for liquid fuels and also address the build-up of greenhouse gases that cause global climate change [2].

Author details

T.P. Basso, T.O. Basso, C.R. Gallo and L.C. Basso
University of São Paulo, "Luiz de Queiroz" College of Agriculture, Brazil

7. References

[1] Gomez LD, Steele-King CG, McQueen-Mason SJ. Sustainable liquid biofuels from biomass: the writing's on the walls. New Phytologist 2008;178:473-485.

[2] Yang B, Wyman CE. Pretreatment: the key to unlocking low-cost cellulosic ethanol. Biofuels, Bioproducts and Biorefining 2008;2:26-40.

[3] Jagger A. Brazil invests in second-generation biofuels. Biofuels, Bioproducts and Biorefining 2009;3:8-10.

[4]   Galbe M, Zacchi G. A review of the production of ethanol from softwood. Applied of
      Microbiology and Biotechnology 2002;59:618-628.
[5]   Pandey A, Socol CR, Nigam P, Soccol VT. Biotechnological potential of agro-industrial
      residues. I: sugarcane bagasse 2000;74:69-80.
[6]   Mielenz JR. Ethanol production from biomass: technology and commercializantion
      status. Current Opinion in Microbiology 2001;4:324-329.
[7]   Cardona CA, Quintero JA, Paz IC. Production of bioethanol from sugarcane bagasse:
      Status and perspectives. Bioresource and Technology 2010;101:4754-4766.
[8]   Kumar S, Singh, SP, Mishra IM, Adhikari DK. Recent advances in production of
      bioethanol from lignocellulosic biomass. Chemistry Engineering Technology
      2009;32(4)517-526. Applied Biochemistry and Biotechnology 2001;91:5–21
[9]   Takashima S, Nakamura A, Hidaka M, Masaki H, Uozumi T. Molecular cloning and
      expression of the novel fungal β-glucosidase genes from *Humicola grisea* and
      Trichoderma reesei. Journal of Biochemistry (Tokyo) 1999;125:728–736.
[10] Zhang YHP, Himmel ME, Mielenz JR. Outlook for cellulase improvement: Screening
      and selection strategies. Biotechnology Advances 2006;24;452-481.
[11] Chen H, Jin S. Effect of ethanol and yeast on cellulose activity and hydrolysis of
      crystalline cellulose. Enzyme and Microbial Technology 2006;39:1430-1432.
[12] Stenberg K, Galbe M, Zacchi G. The influence of lactic acid formation on the
      simultaneous saccharification and fermentation (SSF) of softwood to ethanol. Enzyme
      Microbiology Technology 2000;26:71–79.
[13] Vasquez PM, Silva JNC, Souza Jr MB, Pereira Jr N. Enzymatic hydrolysis optimization
      to ethanol production by simultaneous saccharification and fermentation. Applied
      Biochemistry and Biotechnology 2007; 136-149:141-153.
[14] Krishnan C, Sousa Lda C, Jin M, Chang L, Dale BE, Balan V. Alkali-based AFEX
      pretreatment for the conversion of sugarcane bagasse and cane leaf residues to ethanol.
      Biotechnology and Bioengineering 2010; 107:441-50.
[15] Foreman PK, Brown D, Dankmeyer L, Dean R, Diener S, Dunn-Coleman NS,
      Goedegebuur F, Houfek TD, England GJ, Kelley AS, Meerman HJ, Mitchell T,
      Mitchinson C, Olivares HA, Teunissen PJM, Yao J, Ward M. Transcriptional regulation
      of biomass-dedrading enzymes in the filamentous fungis Trichoderma reesei. The
      Journal of Biological Chemistry 2003;278(34)31988-31997.

Chapter 8

Methanogenic System of a Small Lowland Stream Sitka, Czech Republic

Martin Rulík, Adam Bednařík, Václav Mach, Lenka Brablcová, Iva Buriánková, Pavlína Badurová and Kristýna Gratzová

Additional information is available at the end of the chapter

1. Introduction

Methane (CH_4) is an atmospheric trace gas present at concentration of about 1.8 ppmv, that represents about 15% of the anthropogenic greenhouse effect (Forster et al. 2007). The atmospheric CH_4 concentration has increased steadily since the beginning of the industrial revolution (~ 0.7 ppmv) and is stabilized at ~1.8 ppmv from 1999 to 2005 (Forster et al. 2007). An unexpected increase in the atmospheric growth of CH_4 during the year 2007 has been recently reported (Rigby et al. 2008), indicating that the sources and sinks of atmospheric CH_4 are dynamics, evolving, and not well understood. Freshwater sediments, including wetlands, rice paddies and lakes, are thought to contribute 40 to 50 % of the annual atmospheric methane flux (Cicerone & Oremland 1988; Conrad 2009).

The river hyporheic zone, volume of saturated sediment beneath and beside streams containing some proportion of water from surface channel, plays a very important role in the processes of self-purification because the river bed sediments are metabolically active and are responsible for retention, storage and mineralization of organic matter transported by the surface water (Hendricks 1993; Jones & Holmes 1996, Baker et al. 1999, Storey et al. 1999, Fischer et al. 2005). The seemingly well-oxygenated hyporheic zone contains anoxic and hypoxic pockets („anaerobic microzones") associated with irregularities in sediment surfaces, small pore spaces or local deposits of organic matter, creating a 'mosaic' structure of various environments, where different microbial populations can live and different microbially mediated processes can occur simultaneously (Baker et al. 1999, Morrice et al. 2000, Fischer et al. 2005). Moreover, hyporheic-surface exchange and subsurface hydrologic flow patterns result in solute gradients that are important in microbial metabolism. Oxidation processes may occur more readily where oxygen is replenished by surface water infiltration, while reduction processes may prevail where surface-water exchange of oxygen is less, and the reducing potential of the environment is greater (Hendricks 1993). As water moves through the hyporheic zone, decomposition of the organic matter consumes oxygen, creating oxygen gradients along the flow path. Thus,compared to marine or lake surface sediments, where numerous studies on O_2 profiles have showed that O_2 concentrations become zero within less than 3 mm from the surface, the hyporheic sediment might be well-oxygenated habitats even up to the depth of 80 cm (e.g. Bretschko 1981, Holmes et al. 1994) . The extent of the oxygen gradient is determined by the interplay between flow path lengh, water velocity, the ratio of surface to ground water, and the amount and quality of organic matter. Organic matter decomposition in sediments is an important process in global and local carbon budgets as it ultimately recycles complex organic compounds from terrestrial and aquatic environments to carbon dioxide and methane. Methane is a major component in the carbon cycle of anaerobic aquatic systems, particularly those with low sulphate concentrations. Since a relatively high production of methane has been measured in river sediments (e.g. Schindler & Krabbenhoft 1998, Hlaváčová et al. 2005, Sanders et al. 2007, Wilcock & Sorrell 2008, Sanz et al. 2011), we proposed that river sediments may act as a considerable source of this greenhouse gas which is important in global warming (Hlaváčová et al. 2006).

Breakdown of organic matter and gas production are both results of well functioned river self-purification. This degrading capacity, however, requires intensive contact of the water with biologically active surfaces. Flow over various morphological features ranging in size from ripples and dunes to meanders and pool-riffle sequences controls such surface-subsurface fluxes. Highly permeable streambeds create opportunities for subsurface retention and long-term storage, and exchange with the surface water is frequent. Thus, study of the methane production within hyporheic zone and its subsequent emission to the atmosphere can be considered as a measure of mineralization of organic matter in the freshwater ecosystem and might be used in evaluation of both the health and environmental quality of the rivers studied.

Methane (CH_4) is mostly produced by methanogenic archaea (Garcia et al. 2000, Chaban et al. 2006) as a final product of anaerobic respiration and fermentation, but there is also aerobic methane formation (e.g. Karl et al. 2008). Methanogenic archaea are ubiquitous in anoxic environments and require an extremely low redox potential to grow. They can be found both in moderate habitats such as rice paddies (Grosskopf et al. 1998a,b), lakes (Jürgens et al. 2000, Keough et al. 2003) and lake sediments (Chan et al. 2005), as well as in the gastrointestinal tract of animals (Lin et al. 1997) and in extreme habitats such as hydrothermal vents (Jeanthon et al. 1999), hypersaline habitats (Mathrani & Boone 1995) and permafrost soils (Kobabe et al. 2004, Ganzert et al. 2006). Rates of methane production and consumption in sediments are controlled by the relative availability of substrates for methanogenesis (especially acetate or hydrogen and carbon dioxide). The most important immediate precursors of methanogenesis are acetate and H_2/CO_2. The acetotrophic methanogens convert acetic acid to CH_4 and CO_2 while the hydrogenotrophic methanogens convert CO_2 with H_2 to CH_4 (Conrad 2007).

Methane oxidation can occur in both aerobic and anaerobic environments; however, these are completely different processes involving different groups of prokaryotes. Aerobic methane oxidation is carried out by aerobic methane oxidizing bacteria (methanotrophs, MOB), while anaerobic methane oxidizers, discovered recently, thrive under anaerobic conditions and use sulphate or nitrate as electron donors for methane oxidation (e.g. Strous & Jetten 2004). MOB are a physiologically specialized group of methylotrophic bacteria capable of utilizing methane as a sole source of carbon and energy, and they have been recognized as major players in local and global elemental cycling in aerobic environments (Hanson & Hanson 1996, Murrell et al. 1998, Costelo & Lidstrom 1999, Costelo et al. 2002, McDonald et al. 2008). Aerobic MOB have been detected in a variety of environments, and in some they represent significant fractions of total microbial communities (e.g. Henckel et al. 1999; Carini et al, 2005, Trotsenko & Khmelenina 2005, Kalyuzhnaya et al. 2006). However, the data on the diversity and activity of methanotrophic communities from the river ecosystems are yet fragmentary. Methanotrophs play an important role in the oxidation of methane in the natural environment, oxidizing methane biologically produced in anaerobic environments by the methanogenic archaea and thereby reducing the amount of methane released into the atmosphere.

The present investigation is a part of a long-term study focused on organic carbon and methane dynamics and microbial communities in hyporheic zone of a Sitka, small lowland stream in Czech Republic. The overall purpose of this research was to characterize spatial distribution of both methanogens and methanotrophs within hyporheic sediments and elucidate the differences in methane pathways and methane production/consumption as well as methane fluxes and atmospheric emissions at different sites along a longitudinal profile of the stream.

2. Material and methods

2.1. Study site

The sampling sites are located on the Sitka stream, Czech Republic (Fig. 1). The Sitka is an undisturbed, third-order, 35 km long lowland stream originating in the Hrubý Jeseník mountains at 650 m above sea level. The catchment area is 118.81km^2, geology being composed mainly of Plio-Pleistocene clastic sediments of lake origin covered by quaternary sediments. The mean annual precipitation of the downstream part of the catchment area varies from 500 to 600 mm. Mean annual discharge is 0.81 m^3.s^{-1}. The Sitka stream flows in its upper reach till Šternberk through a forested area with a low intensity of anthropogenic effects, while the lower course of the stream naturally meanders through an intensively managed agricultural landscape. Except for short stretches, the Sitka stream is unregulated with well-established riparian vegetation. River bed sediments are composed of gravels in the upper parts of the stream (median grain size 13 mm) while the lower part, several kilometres away from the confluence, is characterised by finer sediment with a median grain size of 2.8 mm. The Sitka stream confluences with the Oskava stream about 5 km north of Olomouc. More detailed characteristics of the geology, gravel bar, longitudinal physicochemical (e.g. temperature, pH, redox, conductivity, O_2, CH_4, NO_3, SO_4) patterns in the sediments and a schematic view of the site with sampling point positions have been published previously (Rulík et al. 2000, Rulík & Spáčil 2004). Earlier measurements of a relatively high production of methane, as well as potential methanogenesis, confirmed the suitability of the field sites for the study of methane cycling (Rulík et al. 2000, Hlaváčová et al. 2005, 2006).

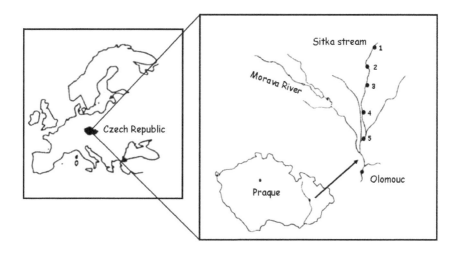

Figure 1. A map showing the location of the Sitka stream. Black circles represents the study sites (1-5)

2.2. Sediment sample collecting and sample processing

Five localities alongside stream profile were chosen for sampling sediment and interstitial water samples based on previous investigations (Figure 2, Table 1). Hyporheic sediments were collected with a freeze-core using N_2 as a coolant (Bretschko & Klemens 1986) throughout summer period 2009-2011. At each locality, three cores were taken for subsequent analyses. After sampling, surface 0-25 cm sediment layer and layer of 25-50 cm in depth were immediately separated and were stored at a low temperature whilst being transported to the laboratory. Just after thawing, wet sediment of each layer was sieved and only particles < 1 mm were considered for the following microbial measurements and for all microbial activity measurements since most of the biofilm is associated with this fraction (Leichtfried 1988).

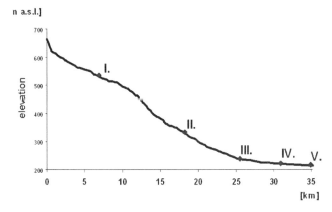

Figure 2. Graphic depiction of the thalweg of the Sitka stream with sampling localities. The main source of pollution is an effluent from Šternberk city sewage water plant, located just in the middle between stretch II and III.

Variable/ Locality	I.	II.	III.	IV.	V.
elevation above sea-level [m]	535	330	240	225	215
distance from the spring [km]	6,9	18,2	25,6	30,9	34,9
channel width [cm]	523	793	672	444	523
average flow velocity [m.s^{-1}]	0,18	0,21	0,46	0,42	0,18
stretch longitude [km]	12,6	9,3	6,3	4,7	2,3
stretch surface area [km^2]	0,043	0,06	0,043	0,024	0,012
stretch surface area (%)	24	32	24	13	7
dominant substrate composition	gravel	gravel	gravel	silt-clay	gravel-sand
grain median size [mm]	12,4	12,9	13,2	0,2	5,4
surface water PO$_4^{3-}$ [mg L^{-1}]	0,15	0,24	7,0	2,6	1,8
surface water N - NO$_3^-$ [mg L^{-1}]	0,01	0,21	1,2	0,5	0,18
surface water N - NH$_4^+$ [mg L^{-1}]	0,39	0,26	0,66	0,72	0,61
surface dissolved oxygen saturation [%]	101,7	110,0	105,8	108,5	103,5
surface water conductivity [µS.cm^{-1}]	107,5	127,5	404,8	394,0	397,7
hyporheic water conductivity [µS.cm^{-1}]	115,3	138,3	414,5	506,5	416,2
surface water temperature [°C]	8,1	9,7	10,7	11,1	8,9
surface water DOC [mg L^{-1}]	2,47	0,81	2,62	2,69	3,74
hyporheic water DOC [mg L^{-1}]	2,05	1,31	2,71	5,76	2,62

Table 1. Longitudinal physicochemical patterns of the Sitka stream (annual means). Hyporheic water means mix of interstitial water taken from the depth 10 up to 50 cm of the sediment depth

A few randomly selected subsamples (1 mL) were used for extraction of bacterial cells and, consequently, for estimations of bacterial numbers; other sub-samples were used for measurement of microbial activity and respiration, organic matter content determination, etc. Sediment organic matter content was determined by oven-drying at 105 ᵒC to constant weight and subsequent combustion at 550 ᵒC for 5 hours to obtain ash-free dry weight (AFDW). Organic matter values were then converted to carbon equivalents assuming 45 % carbon content of organic matter (Meyer et al. 1981). Sediment from another freeze-core was oven-dried at 105 °C and subjected to granulometric analysis. Grain size distribution and descriptive sediment parameters were computed using the database SeDi (Schönbauer & Lewandowski 1999).

2.3. Water samples and analysis of methane

Surface water was collected from running water at a depth of 10 cm below the surface level in autumn 2009 at each study site. Interstitial water samples were collected using a set of 5–6 minipiezometers (Trulleyová et al. 2003) placed at a depth of about 20-50 cm randomly in sediments at each study site. The initial 50–100 mL of water was used as a rinse and discarded. As usual, two subsamples of interstitial water from each minipiezometer were collected from a continuous column of water with a 100 mL polypropylene syringe connected to a hard PVC tube, drawn from a minipiezometer and injected into sterile, clear vials (40 mL) with screw-tops, covered by a polypropylene cap with PTFE silicone septa (for analysis of dissolved gasses) and stored before returning to the laboratory. All samples were taken in the morning between 9 a.m. and 12 noon. All measurements were done during the normal discharge levels (i.e. no spates or high flood levels were included). Interstitial water temperature, dissolved oxygen (mg L^{-1} and percent saturation) and conductivity were measured in the field with a portable Hanna HI 9828 pH/ORP/EC/DO meter. Dissolved organic carbon (DOC) was measured by Pt-catalysed high temperature combustion on a TOC FORMACS[HT] analyser. Long term observation of interstitial water temperature was carried out using temperature dataloggers Minikin (EMS Brno, Czech Republic) buried in the sediment depth of 25-30 cm for a period of one year. Dissolved ferrous iron (Fe^{2+}) concentration was measured using absorption spectrophotometry after reaction with 1,10-phenanthroline. Concentrations of organic acids were meausred using capilary electrophoresis equipped with diode array detector HP 3D CE Agilent (Waldbron, Germany). Limits of detection for particular organic acids were set as following: LOD (acetate) = 6,2 μmol L^{-1}; LOD (propionate) = 4,8 μmol L^{-1}; LOD (butyrate) = 2,9 μmol L^{-1}; LOD 32 (valerate) = 1,8 μmol L^{-1}.

Concentrations of dissolved methane in the stream and interstitial water were measured directly using a headspace equilibration technique. Dissolved methane was extracted from the water by replacing 10 mL of water with N_2 and then vigorously shaking the vials for 15 seconds (to release the supersaturated gas from the water to facilitate equilibration between the water and gas phases). All samples were equilibrated with air at laboratory temperature. Methane was analysed from the headspace of the vials by injecting 2ml of air sub-sample with a gas-tight syringe into a CHROM 5 gas chromatograph, equipped with the flame ionization detector (CH4 detection limit = 1μg L^{-1}) and with the 1.2m PORAPAK Q column (i.d. 3 mm), with nitrogen as a carrier gas. Gas concentration in water was calculated using

Henry's law. The saturation ratio (R) was calculated as the measured concentration of gas divided by the concentration in equilibrium with the atmosphere at the temperature of the water sample using the solubility data of Wiesenburg & Guinasso (1979).

2.4. Methanogenic potential and methanotrophic activity

The rate of methane production (methanogenesis) was measured using the PMP method (Segers 1998). C-amended solutions (flushed for 5 minutes with N_2) with acetate $Ca(CH_3COO)_2$ (100 mg C in the incubation flask) were used for the examination of methanogenic potential. All laboratory sediment incubations were performed in 250-mL dark glass flasks, capped with rubber stoppers, using approximately 100 g (wet mass) of sediment (grain size < 1 mm) and 180 mL of amended solution or distilled water. The headspace was maintained at 20 mL. Typically, triplicate live and dead (methanogenesis was inhibited by addition of 1.0 mM chloroform) samples from each depth were stored at 20°C in the dark and the incubation time was 72 hours; however, subsamples from the headspace atmosphere were taken every 24 hours. Gas production was calculated from the difference between final and initial headspace concentration and volume of the flask; results are expressed per volume unit of wet sediment ($CH4$ mL^{-1} WW $hour^{-1}$) or per unit dry weight of sediment per one day (μg CH_4 kg^{-1} DW day^{-1}). Rate of potential methane oxidation (methanotrophy) was measured using modified method of methane oxidation in soil samples from Hanson (1998). Briefly, 50 mL of methane was added by syringe to the closed incubation flask with the sieved sediment and then the pressure was balanced to atmospheric pressure. All laboratory sediment incubations were performed in 250-mL dark glass flasks, capped with rubber stoppers, using approximately 100 g (wet mass) of sediment (grain size < 2 mm). Typically, triplicate live and dead (samples killed by $HgCl_2$ to arrest all biological activity) samples from each depth were stored at 20°C in the dark, and incubation time was 72 hours; however, subsamples from the headspace atmosphere were taken every 24 hours. Potential CH_4 oxidation rates at the different concentrations were obtained from the slope of the CH_4 decrease with time ($r^2 > 0.90$; methane oxidation was calculated from the difference between final and initial headspace concentration and volume of the flask; results are expressed per volume unit of wet sediment ($CH4$ mL^{-1} WW $hour^{-1}$) or per unit dry weight of sediment per one day (mg CH_4 kg^{-1} DW day^{-1}).

2.5. Fluxes of methane across the sediment-water interface

Fluxes of methane across the sediment-water interface were estimated either by direct measurement with benthic chambers or calculated by applying Fick's first law.

Benthic fluxes

The methane fluxes across the sediment-water interface were measured using the method of benthic chambers (e.g. Sansone et al. 1998). Fluxes were measured during the summer months (VII, VIII, IX). The plexiglas chamber (2.6 dm^3) covered an area 0.0154 m^2. The chambers (n = 7) were installed randomly and gently anchored on the substrate without disturbing the sediment. Samples to determine of initial concentration of CH_4 were collected from each chamber before the beginning of incubation. Incubation time was 24 hours. Samples of water were stored in 40 ml glass vials closed by cap with PTFE/silicone septum until analysis.

Diffusive fluxes

Fluxes of methane between the sediment and overlying water were calculated from Fick's first law as described by Berner (1980):

$$J = -D_S \times \Phi \times \left(\Delta C / \Delta x\right) \tag{1}$$

where J is the diffusive flux in μg m^{-2} s^{-1}, Φ is the porosity of the sediment, D_S is the bulk sediment diffusion coefficient in cm^{-2} s^{-1}, $\Delta C/\Delta x$ is the methane concetrations gradient in μg cm^{-3} cm^{-1}. Bulk sediment diffusion coefficient (D_S) is based on diffusion coefficient for methane in the water (D_0) and tortuosity (θ) according to the formula:

$$D_S = D_0\theta^{-2} \tag{2}$$

Tortuosity (θ) is possible calculate from porosity according to equation (Boudreau 1996):

$$\theta^{-2} = 1 - ln\left(\Phi^2\right) \tag{3}$$

Diffusive fluxes of CH_4 were determined at all five study sites along the longitudinal profile of the Sitka stream.

2.6. Measurement of emissions

Gas flux across the air-water interface was determined by the floating chamber method four times during the year period in 2005 – 2006. The open-bottom floating PE chambers (5L domes with an area of 0.03 m^2) were maintained on the water's surface by a floating body (Styrene) attached to the outside. The chambers (n = 4 – 5) were allowed to float on the water's surface for a period of 3 hours. Previous measurements confirmed that time to be quite enough to establish linear dependence of concentration change inside the chambers on time for the gas samples collected every 30 min over a 3 hour period. Due to trees on the banks, the chambers at all study sites were continuously in the shade. On each sampling occasion, ambient air samples were collected for determining the initial background concentrations. Samples of headspace gas were collected through the rubber stopper inserted at the chamber's top, and stored in 100mL PE gas-tight syringes until analysis. Emissions were calculated as the difference between initial background and final concentration in the chamber headspace, and expressed on the 1m^2 area of the surface level per day according to the formula:

$$F = \left[\left(c_I - c_R\right) * V * 24 / t * 1000\right] / p \tag{4}$$

where F is a gas flux in mg m^{-2}day^{-1}; c_i is a concentration of particular gas in the chamber headspace in µg L^{-1}; c_R is a concentration of particular gas in background air; V is volume of the chamber in L; t is time of incubation in hr; p is an area of chamber expressed in m^2. For each chamber, the fluxes were calculated using linear regression based on the concentration change as a function of time, regardless of the value of the coefficient of determination (cf. Duchemin et al. 1999, Silvenoinen et al. 2008).

In order to assess emissions produced from a total stream area, the stream was divided into five stretches according to the channel width, water velocity and substrate composition. For each stretch we have then chosen one representative sampling site (locality I-V) where samples of both stream and interstitial waters and sediments, respectively, were repeatedly taken. Localities were chosen in respect to their character and availability by car and measuring equipments. For calculation of whole-stream gases emissions into the atmosphere, the total stream area was derived from summing of 14 partial stretches. The area of these stretches was caculated from known lenght and mean channel width (measured by a metal measuring type). Longitudinal distance among the stretches was evaluated by using ArcGIS software and GPS coordinates that have been obtained during the field measurement and from digitalised map of the Sitka stream. The total area of the Sitka stream was estimated to be 181 380 m^2 or 0.18 km^2. Stretches have differed in their percentual contribution to this total area and also by their total lenght (Table 1).

The total annual methane emissions to the atmosphere from the five segments of the Sitka stream, E_a (kg yr^{-1}) were derived from seasonal average, maximum or minimum emissions measured on every locality and extrapolated to the total area of the particular segment. The total methane emissions produced by the Sitka stream annualy were then calculated according to the following formula:

$$E_a = \left(\Sigma \ p_i * F_i * 365 \right) / 1000000 \qquad (5)$$

where E_a is average, maximal or minimal assess of emission of methane from the total stream area in kilograms per year; p_i is an area of stretch (in m^2) representing given locality; F_i is average, maximal or minimal assess of the methane from a given locality expressed in mg m^{-2}day^{-1}.

2.7. Carbon isotopic composition of dissolved methane and carbon dioxide in sediments

Interstitial water samples for carbon isotopic analysis of methane and carbon dioxide were collected in 2010 - 2011 through three courses at study site. Sampling was performed by set of minipiezometers placed in a depth of 20 to 60 cm randomly in a sediment. After sampling, refrigerated samples were transported (within 72 hours) in 250 mL bottles to laboratory at the Department of Plant Physiology, Faculty of Science University of South Bohemia in Ceske Budejovice, which are equipped with mass spectrometry for carbon isotopes measurements. Firstly both water samples, for methane and for carbon dioxide, were extracted to helium headspace. After relaxation time isotopic equilibrium was achieved and four subsamples of gas were determined by GasBanch (ThermoScientific) and IRMS Delta^{plus}XL equiped by TC/EA (ThermoFinnigan) for analysis of $\delta^{13}CO_2$. Afterwards $\delta^{13}CO_2$ of water samples were calculated from gaseous $\delta^{13}CO_2$ by fractionation factor from a linear equation (Szaran 1997):

$$\varepsilon_\infty{}^{13}C = -\left(0.0954 \pm 0.0027\right) T[°C] + (10.41 \pm 0.12) \qquad (6)$$

Stable isotope analysis of $^{13}C/^{12}C$ in gas samples was performed using preconcentration, kryoseparation of CO_2 and gas chromatograph combustion of CH_4 in PreCon (ThermoFinnigan) coupled to isotope ratio mass spectrometer (IRMS, Delta Plus XL, ThermoFinnigan, Brehmen, Germany). After conversion of CH_4 to CO_2 in the Finningan standard GC Combustion interface CO_2 will be tranfered into IRMS. The obtained $^{13}C/^{12}C$ ratios (R) will be referenced to $^{13}C/^{12}C$ of standard V-PDB (Vienna-Pee-Dee Belemnite)(Rs), and expressed as $\delta^{13}C$ = (R$_{sample}$/R$_{standard}$ − 1) x 1000 in ‰. The standard deviation of $\delta^{13}C$ determination in standard samples is lower than 0.1‰ with our instrumentation. From our data, we also calculated an apparent fractionation factor α_C that is defined by the measured δCH_4 and δCO_2 (Whiticar et al. 1986):

$$\alpha_C = \left(\delta CO_2 + 10^3\right) / \left(\delta CH_4 + 10^3\right) \qquad (7)$$

This fractionation factor gives rough idea of magnitude of acetoclastic and hydrogenotrophic methanogenesis.

2.8. Abundance of microbial cells and microbial community composition

For measuring of microbial parameters, formaldehyde fixed samples (2 % final conc.) were first mildly sonicated for 30 seconds at the 15 % power (sonotroda MS 73, Sonopuls HD2200, Sonorex, Germany), followed by incubation for 3 hours under mild agitation with 10 mL of detergent mixture (Tween 20 0.5%, vol/vol, tetrasodium pyrophosphate 0.1 M and distilled water) and density centrifugation (Santos Furtado & Casper 2000, Amalfitano & Fazi 2008). For density centrifugaton, the non-ionic medium Nycodenz (1.31 g mL^{-1}; Axis- Shield, Oslo, Norway) was used at 4600 G for 60 minutes (Rotofix 32A, Hettich, Germany). After the preparation processes, a 1 mL of Nycodenz was placed underneath 2 ml of treated slurry using a syringe needle (Fazi et al. 2005). 1 ml of supernatant was then taken for subsequent analysis.

2.9. Total cell numbers (TCN)

The supernatant was filtered onto membrane filters (0.2 μm GTTP; Millipore Germany), stained for 10 minutes in cold and in the dark with DAPI solution (1 mg/ ml; wt/ vol; Sigma, Germany) and gently rinsed in distilled water and 80 % ethanol. Filters were air-dried and fixed in immersion oil. Stained cells were enumerated on an epifluorescence microscope (Olympus BX 60) equipped with a camera (Olympus DP 12) and image analysis software (NIS Elements; Laboratory Imaging, Prague, Czech Republic). At least 200 cells within at least 20 microscopic fields were counted in three replicates from each locality. TCN was expressed as bacterial numbers per 1 mL of wet sediments.

2.10. Procaryotic community composition

The methanogenic archaea, three selected methanogen families (*Methanobacteriaceae*, *Methanosetaceae* and *Methanosarcinaceae*) and methanotrophic bacteria belonging to groups I and II were detected using FISH (Fluorescence in situ hybridization) with 16S rRNA-targeted oligonucleotide probe labelled with indocarbocyanine dye Cy3. The prokaryotes were hybridized according to the protocol by Pernthaler et al. (2001). Briefly, the supernatants which were used also for TCN were filtered onto polycarbonate membrane filters (0.2 μm GTTP; Millipore), filters were cut into sections and placed on glass slides. For the hybridization mixtures, 2 μL of probe-working solution was added to 16 μL of hybridization buffer in a microfuge tube. Hybridization mix was added to the samples and the slides with filter sections were incubated at 46 °C for 3 hours. After incubation, the sections were transferred into preheated washing buffer (48 °C) and incubated for 15 minutes in a water bath at the same temperature. The filter sections were washed and air-dried. The DAPI staining procedure followed as previously described. Finally, the samples were mounted in a 4:1 mix of Citifluor and Vecta Shield. The methanogens and methanotrophs were counted in three replicates from each locality and the relative proportion of bacteria, archaea, methanogens and methanotrophs to the total number of DAPI stained cells was then calculated.

2.11. Nucleic acid extraction and Denaturing gradient gel electrophoresis (DGGE)

Nucleic acids were extracted from 0,3 g of sieved sediment with a Power Soil DNA isolation kit (MoBio, Carlsbad, USA) according to the manufacturer's instructions. 16S rRNA gene fragments (~350 bp) were amplified by PCR using primer pair specific for methanogens. Primer sequences are as follows, 0357 F-GC 5'-CCC TAC GGG GCG CAG CAG-3' (GC clamp at 5'-end CGC CCG CCG CGC GCG GCG GGCGGG GCG GGG GCA CGG GGG G) and 0691 R 5'- GGA TTA CAR GAT TTC AC -3' (Watanabe et al. 2004). PCR amplification was carried out in 50 μL reaction mixture contained within 0.2 mL, thin walled micro-tubes. Amplification was performed in a TC-XP thermal cycler (Bioer Technology, Hangzhou, China). The reaction mixture contained 5 μL of 10 × PCR amplification buffer, 200 μM of each dNTP, 0,8 μM of each primer, 8 μL of template DNA and 5.0 U of FastStart Taq DNA polymerase (Polymerase dNTPack; Roche, Germany). The initial enzyme activation and DNA denaturation were performed for 6 min at 95°C, followed by 35 cycles of 1 min at 95°C, 1 min at 55°C and 2 min at 69°C and a final extension at 69°C for 8 min (protocol by Watanabe et al. 2004). PCR products were visualised by electrophoresis in ethidium bromide stained, 1.5% (w/v) agarose gel.

DGGE was performed with an INGENYphorU System (Ingeny, Netherlands). PCR products were loaded onto a 7% (w/v) polyacrylamide gel (acrylamide: bisacrylamide, 37.5:1). The polyacrylamide gels were made of 0.05% (v/v) TEMED (N,N,N,N-tetramethylenediamine), 0.06% (w/v) ammonium persulfate, 7 M (w/v) urea and 40 % (v/v) formamide. Denaturing gradients ranged from 45 to 60%. Electrophoresis was performed in 1×TAE buffer (40 mM Tris, 1 mM acetic acid, 1 mM EDTA, pH 7.45) and run initially at 110V for 10 min at 60°C, afterwatds for 16 h at 85 V. After electrophoresis, the gels were stained for 60 min with SYBR Green I nucleic acid gel stain (1:10 000 dilution) (Lonza, Rockland USA) DGGE gel was then photographed under UV transilluminator (Molecular Dynamics). Images were arranged by Image analysis (NIS Elements, Czech Republic). A binary matrix was created from the gel image by scoring of the presence or absence of each bend and then the cluster tree was constructed (programme GEL2k; Svein Norland, Dept. Of Biology, University of Bergen).

2.12. PCR amplification, cloning and sequencing of methyl coenzyme M reductase (mcrA) gene

Fragments of the methanogen DNA (~470 bp) were amplified by PCR using *mcrA* gene specific primers. Primer sequences for mcrA gene are as follows, mcrA F 5'-GGTGGTGTACGGATTCACACAAGTACTGCATACAGC-3',mcrA   R   5'-TTCATTGCAGTAGTTATGGAGTAGTT-3'. PCR amplification was carried out in 50 μl reaction mixture contained within 0.2 mL thin walled micro-tubes. Amplification was performed in a TC-XP thermal cycler (Bioer Technology, Hangzhou, China). The reaction mixture contained 5 μL of 10 x PCR amplification buffer, 200 μM of each dNTP, 0.8 μM of each primer, 2 μL of template DNA and 2.5 U of FastStart Taq DNA polymerase (Polymerase dNTPack; Roche, Mannheim, Germany). The initial enzyme activation and DNA denaturation were performed for 6 min at 95°C, followed by 5 cycles of 30s at 95°C, 30s at 55°C and 30s at 72°C, and the temperature ramp rate between the annealing and extension segment was set to 0.1°C/s because of the degeneracy of the primers. After this, the ramp rate was set to 1°C/s, and 30 cycles were performed with the following conditions: 30 s at 95°C, 30 s at 55°C, 30s at 72°C and a final extension at 72°C for 8 min. PCR products were visualised by electrophoresis in ethidium bromide stained, 1.5% (w/v) agarose gel.

Purified PCR amplicons (PCR purification kit; Qiagen, Venlo, Netherlands) were ligated into TOPO TA cloning vectors and transformed into chemically competent *Escherichia coli* TOP10F′ cells according to the manufacturer's instructions (Invitrogen, Carlsbad, USA). Positive colonies were screened by PCR amplification with the primer set and PCR conditions described above. Plasmids were extracted using UltraClean 6 Minute Plasmid Prep Kit (MoBio, Carlsbad, USA), and nucleotide sequences of cloned genes were determined by sequencing with M13 primers in Macrogen company (Seoul, Korea). Raw sequences obtained after sequencing were BLAST analysed to search for the sequence identity between other methanogen sequences available in the GenBank database. Then these sequences were aligned by using CLUSTAL W in order to remove any similar sequences. The most appropriate substitution model for maximum likelihood analysis was identified by Bayesian Information Criterion implemented in MEGA 5.05 software. The phylogenetic tree was constructed by the maximum likelihood method   (Kimura 2-parameter model). The tree topology was statistically evaluated by 1000 bootstrap replicates (maximum likelihood) and 2000 bootstrap replicates (neighbour joining).

3. Results

3.1. Sediment and interstitial water

The physicochemical sediment and interstitial water properties of the investigated sites showed large horizontal and vertical gradients. Sediment grain median size decreased along a longitudinal profile while organic carbon content in a sediment fraction < 1 mm remained unchaged (Table 2). Generally, interstitial water revealed relatively high dissolved oxygen saturation with the exceptions of localities IV and V where concentration of dissolved oxygen sharply decreased with the depth, however, never dropped below ~ 10%. Vice versa, these two localities were characterized by much higher concentrations of ferrous iron and dissolved methane (Table 2) compared to those sites located upstream. Concentration of the ferrous iron reflects anaerobic conditions of the sediment and showed the highest concentration to occur in the deepest sediment layers (40-50cm). Average annual temperatures of interstitial water at localities in downstream part of the Sitka stream were about 2.5 °C higher compared to localities upstream and may probably promote higher methane production occuring here. Precursors of methanogenesis, acetate, propionate and butyrate were found to be present in the interstitial water at all study sites, however, only acetate was measured regularly at higher concentration with maximum concentration reached usually during a summer period.

Variable/ Locality	I	II	III	IV	V
particulate organic C in sediment < 1 mm [%]	0.9	0.9	0.6	1.3	0.7
interstitial dissolved O^2 saturation [%]	80.5	88.1	82.3	38.5	50.9
ferrous iron [mg L^{-1}]	< 1	< 1	1.8	8.1	4.2
acetate [mmol L^{-1}]	0.21	0.34	0.52	1.87	0.29
interstitial CH4 concentration [$\mu g\ L^{-1}$]	4.9	0.7	8.1	2 480.2	42.8
methanogenic potential [pM CH_4 mL^{-1} WW hour^{-1}]	6.6	1.9	2.9	80.7	9.7
methanotrophic activity [nM CH_4 mL^{-1} WW hour^{-1}]	0.3	1.3	28.5	30.3	25.1
average daily interstitial water temperature [°C]	8.7	9.4	11.6	11.2	11.4

Table 2. Selected physicochemical parameters (annual means) of the hyporheic interstitial water and sediments of studied localities taken from the depth 25-30 cm.

3.2. Methanogenic potential and methanotrophic activity of sediments

Methanogenic potential (MP) was found to be significantly higher in the upper sediment layer compared to that from deeper sediment layer. Generally, average MP varied between 0.74-158.6 pM CH_4 mL^{-1} WW hour^{-1} with the highest values found at site IV. Average methanotrophic activity (MA) varied between 0.02– 31.3 nM CH_4 mL^{-1} WW $hour^{-1}$ and the highest values were found to be at the downstream localities while sediment from sites located upstream showed much lower or even negative activity. Similar to MP, values of MA were significantly higher in sediments from upper layers compared to those from deeper layers (e.g. Figs. 3c, 3d).

3.3. Methane concentration along the longitudinal profile, vertical and temporal pattern, stable isotopes

Methane concentrations ranged between 0.18 – 35.47 µg L^{-1} in surface water and showed no expected trend of gradual increase from upstream localities to those laying downstream. However, significant enhancement of CH_4 concentration was found on locality IV and V, respectively. Concentrations of dissolved CH_4 inboth surface and interstitial waters peaked usually during summer and autumn period (Hlaváčová et al. 2005, Mach et al. in review).

Generally, methane concentrations measured in interstitial water were much higher compared to those from surface stream water and on a long-term basis ranged between 0.19 - 11 698.9 µg L^{-1}. Due to low methane concentrations in interstitial water at localities I and II, vertical distribution of its concentrations was studied only at the downstream located sites III-V. Significant increase of the methane with the sediment depth was observed at the localities IV and V, respectively. Namely locality IV proved to be a methane pool, methane concentrations in a depth of 40 cm were found to be one order of magnitude greater than those from the depth of 20 cm (Tab. 3). Recent data from locality IV show much lower methane concentrations in the upper sediment horizons compared to those from deeper layers (Fig. 3a). Considerable lowering of methane concentration in upper sediment horizons is likely caused by oxidizing activity of methanotrophic bacteria (Fig. 3d). while dissolved oxygen concentration sharply decreased with the sediment depth (Fig. 3b).

Locality	Profile (depth)	CH_4 [µg L^{-1}]
III.	Surface water	1.8
	Interstitial water (depth 20cm)	1.44
	Interstitial water (depth 40 cm)	1.52
IV.	Surface water	5.52
	Interstitial water (depth 20 cm)	1 523.9
	Interstitial water (depth 40 cm)	11 390.54
V.	Surface water	4.72
	Interstitial water (depth 20 cm)	6.92
	Interstitial water (depth 40 cm)	24.4

Table 3. Average concentrations of methane in the vertical sediment profile at localities III-V compared to those from surface water at the same sites

Usually, both the surface and interstitial water were found to be supersaturated compared to the atmosphere with locality IV displaying saturation ratio R to be almost 195 000. This high supersaturation greatly promote diffusive fluxes of methane to the atmosphere across air-water interface and is also an important mechanisms for loss of water column CH_4.

Stable carbon isotope signature of carbon dioxide ($\delta^{13}C$-CO_2) measured in the interstitial water ranged from -19.8 ‰ to -0.8 ‰, while carbon isotope signature of methane ($\delta^{13}C$-CH_4) ranged between -72 ‰ to -19.8 ‰. This relatively high variation in the methane isotopic values could be caused due to consequential fractionation effects preferring light carbon isotopes like methane oxidation or fractionation through diffusion and through flow of an interstitial water. Contrary, the narrow range of the $\delta^{13}C$-CH_4 was found in the sediment depth of 40-60 cm where a high methane production has occured. Here, the $\delta^{13}C$-CH_4 values varied only from -67.9 ‰ to -72 ‰. Apparent fractionation factor (αc) varied also greatly from 1,004 to 1,076. Usually values of $\alpha c > 1.065$ and $\alpha c < 1.055$ are characteristic for environments dominated by hydrogenothropic and acetoclastic methanogenesis, respectively. Our measurements indicate predominant occurrence of a hydrogenothropic methanogenesis in the high methanogenic zones where the most amount of methane is produced and $\delta^{13}C$ of CO_2 values were markedly depleted (i.e. ^{13}C enriched). This could be caused by enhanced carbon dioxide consumption by hydrogenothrophic methanogens, strongly preferring light isotopes. Nevertheless, both acetoclastic and hydrogenotrophic pathways take part in the methanogenesis along the longitudinal profile of the Sitka stream.

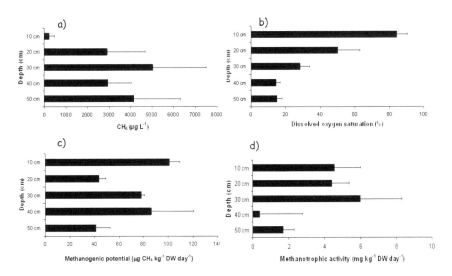

Figure 3. Vertical distribution of methane concentration in the interstitial water at study site IV, horizontal bars indicate 1 SE

3.4. Fluxes of methane across the sediment-water and the air-water interfaces

Methane diffusion rate from deeper sediment layers depends on a methane concentration gradient whilst is affected by oxidation and rate of methanotrophic bacteria consumption. When diffusion fluxes are positive (positive values indicate net CH_4 production), then surface water is enriched by methane which in turn may be a part of downstream transport or is further emitted to the atmosphere (Fig. 4).

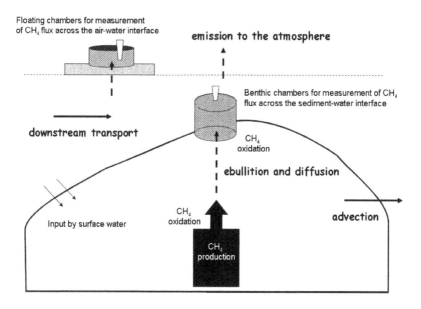

Figure 4. Possible fate of the methane within hyporheic zone and two kinds of chambers for measurement of methane fluxes. Providing that some sites along the longitudinal stream profile should be sources of methane for the stream water, we chose locality IV to be suitable for benthic fluxes measurements.

On the contrary, when the fluxes of methane across the sediment-water interface are negative then all methane produced in the sediments is likely oxidized and consumed by methanotrophic bacteria here or transported via subsurface hyporheic flow.

Calculated diffusive fluxes of CH_4 ranged from 0.03 to 2307.32 µg m^{-2} day^{-1} along the longitudinal profile. The lowest average values of diffusive fluxes were observed at study site II (0.11 ± 0.05 µg m^{-2} day^{-1}) while the highest average values were those observed at study site IV (885.81 ± 697.54 µg m^{-2} day^{-1}). Direct benthic fluxes of CH_4 using the benthic chambers were measured at study site IV only and ranged from 0.19 to 82.17 mg m^{-2} day^{-1}. We observed clear negative relationships between benthic methane fluxes and the flow discharge. During higher discharges when the stream water is pushed into sediments, methane diffusing from deeper sediments upward is either transported by advection through sediments downstream or is probably almost completely oxidized by methanotrophic bacteria due to increasing oxygen supply from the surface stream. As a consequence, very low or no benthic fluxes were recorded during the time of high flow discharge. Compared to calculated diffusive fluxes it is clear that fluxes obtained by direct measurement were approximately 15× higher than the fluxes calculated with using Fick's first law. Thus, direct benthic fluxes were used for a calculation of water column CH_4 budget.

Gaseous fluxes from surface water to the atmosphere were found at all localities except locality I, where emissions were not mesured directly but were calculated lately using a known relationships between concentrations of gases in surface water and their emissions to the atmosphere found at downstream laying localities II-V. Methane showed an increase in emissions toward downstream where highest surface water concentrations have also occured (Table 4). Methane emissions measured at localities II-V ranged from 0 – 167.35 mg $m^{-2}day^{-1}$ and no gradual increase in downstream end was found in spite of our expectation. However, sharp increase in the amount of methane emitted from the surface water was measured at lowermost localities IV and V (Tab. 4). We found positive, but weak correlation between surface water methane concentrations and measured emissions ($r_s = 0.45$, p < 0.05)(Fig. 5).

Locality/Gas	CH_4 [mg $m^{-2}day^{-1}$]
Locality I.	2.39
Locality II.	0.25 (0 – 0.6) $n = 9$
locality III.	1.3 (0 – 5.01) $n = 10$
Locality IV.	32.1 (7.3 – 87.9) $n = 8$
Locality V.	36.3 (2.8 – 167.4) $n = 12$

Table 4. Average emissions to the atmosphere and their range in parenthesis and from all localities except locality I. Emissions values for the locality I were calculated using a known relationships between concentrations of methane gas in surface water and its emissions to the atmosphere found at downstream laying localities II-V. n means sample size

3.5. Whole-stream emissions E_a

Depending on the time of year we measured the emissions, values of E_a ranged from 430 to 925 kg year^{-1} for methane. Annually, approximately 0.7 tonne of methane was emitted to the atmosphere from the water level of the Sitka stream (total area ca 0.2 km^2). The majority of annual methane emissions (90 %) occured in the lower 7 km of the stream (stretch IV and V) that represents only 1/5 of the total stream area. In addition, contribution of methane emissions to the total annual emissions was found to be the highest during spring-summer period (Mach et al. in review).

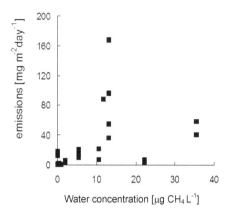

Figure 5. Relationships between atmospheric emissions and surface water concentrations of the methane. Each point represents the mean of five replicate emission measurements and the two replicates of stream water methane concentrations at all

3.6. Sitka stream water column CH$_4$ budget for the experimental stretch of a stream

The potentially important source and sinks terms for dissolved methane in the water column of the Sitka stream are shown in Figure 6. Previously calculated rates of inputs (benthic fluxes) and loss of dissolved CH$_4$ through evasion to the atmosphere can be combined together with advection inputs and losses to yield a CH$_4$ dynamics (budget) for any particular section of the stream.

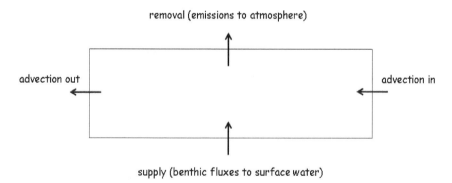

Figure 6. Simple box model used to calculate a CH$_4$ budget for the Sitke stream experimental section; advection in + supply = advection out + removal (box adjusted after de Angelis & Scranton 1993)

The CH$_4$ budget determined for the 2011 sampling period in an experimental stream section is summarized in Figure 7. Benthic fluxes were measured along a stream section 45 m long with an area being ~ 200 m². Positive fluxes of CH₄ were found to occur at 30.9 % of the study area. Assuming that average benthic flux of methane across the sediment-water interface was 15.40 mg m⁻² day⁻¹, the benthic flux of 3081.39 mg CH₄ day⁻¹ should occur from the whole area of 200 m². Average emission flux of CH₄ across the water-air interface for all study sites was determined to be 14.47 ± 4.73 mg CH₄ m⁻² day⁻¹. This value is slightly lower than the direct benthic flux of CH₄ and suggests that some portion of methane released from the bottom sediments may contribute to increasing concentration of CH₄ in the surface water. Average flow of the Sitka stream during time of benthic fluxes measurements was 0.351 m³s⁻¹ (i.e. 351 L⁻¹s⁻¹). Therefore, we may expect that water column was enriched at least by 187.4 mg (i.e. 0.006 µg L⁻¹) of CH₄ from sediment at 45 m long section near study site IV during one day. Next study site V is located some 4 km downstream from the site IV. Average CH₄ concentration difference in the stream water between these study sites was found to be 3.2 µg L⁻¹ of CH₄ indicating that CH₄ supply exceeds slightly CH₄ removal. Methane fluxes from the sediment would contribute to this concentration difference only by 0.6 µg L⁻¹, thus, the immediate difference in the CH₄ budget found between two studied sites IV and V indicates that there must likely be other sources of methane supply to the stream water (Fig. 7). This „missing source" seems to be relatively small (0.9 mg CH₄ 0.351 m⁻³s⁻¹), however, net accumulation of CH₄ in the stream water during 4 km section of the Sitka stream below study site IV was almost 78 g CH₄ per one day.

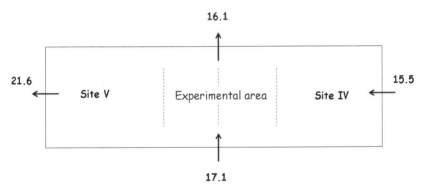

Figure 7. CH₄ budget in mol day⁻¹ for a section of the Sitka stream between study sites IV and V (lenght ca 4 km). The arrows correspond to those depicted in Figure 6.

3.7. Fluorescence in situ hybridization (FISH)

Both methanogenic archaea and aerobic methanotrophs were found at all localities along the longitudinal stream profile. The proportion of these groups to the DAPI-stained cells was quite consistent and varied only slightly but a higher proportion to the DAPI-stained cells in deeper sediment layer 25-50 cm was observed. On average 23,4 % of DAPI-stained cells were detected by FISH with a probe for methanogens while type I methanotrophs reached ~ 21,4 % and type II methanotrophs 11,9 %, respectively. All three groups also revealed non-significant higher proportion to the TCN in deeper sediment layer; the abundance of methanogens and methanotrophs remained almost unchanged with increasing sediment depth. The average abundance of methanogens (0.88 ± 0.28 and $1.07 \pm 0.23 \times 10^6$ cells mL^{-1} in the upper and deeper layer, respectively) and type II methanotrophs ($0.44 \pm 0.14 \times 10^6$ cells mL^{-1} and $0.56 \pm 0.1 \times 10^6$ cells mL^{-1}) increased slightly with the sediment depth , while type I methanotrophs revealed average abundance $0.98 \pm 0.23 \times 10^6$ cells mL^{-1} in the deeper layer being lower compared to abundance $1.07 \pm 0.28 \times 10^6$ cells mL^{-1} found in upper sediment layer (Buriánková et al. 2012). Very recently, however, using the FISH method we found that abundance of methanogens belonging to three selected families reached their maximum in the sediment depth of 20-30 cm and had closely reflected vertical distribution of acetate concentrations. Species of family *Methanobacteriaceae* grow only with hydrogen, formate and alcohols (except methanol), *Methanosarcinaceae* can grow with all methanogenic substrates except formate, and members of *Methanosaetaceae* grow ecxlusively with acetate as energy source. All three families also showed similar proportion to the DAPI stained cells, ranging in average (depth 10-50 cm) from 9.9% (*Methanosarcinaceae*) to 12.3% (*Methanobacteriaceae*) (Fig. 8).

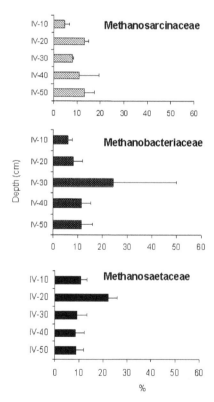

Figure 8. The percentage of chosen methanogenic families as compared to the total bacterial cell numbers found in different sediment layers at locality no. IV, horizontal bars indicate 1 SE

3.8. Denaturing gradient gel electrophoresis and cloning

Methanogenic communities associated with hyporheic sediments at two different depths (0-25 cm and 25-50 cm) along the longitudinal stream profile were compared based on the DGGE patterns. As shown in Fig. 9, the DGGE patterns varied highly among study localities (Fig. 9A), irrespective of the depth (Fig. 9B). However, presence of the bands in all samples indicates that methanogens may occur up to 50 cm of the sediment depth. The number of DGGE bands of the methanogenic archaeal communities was compared either among localites or among different sediment depths. A total of 22 different bands were observed in the DGGE image ranging from 4 (locality II) to 16 (locality IV) in the samples (Fig. 9A).

The number of DGGE bands also ranged from 2 to 10 for the samples from upper layer (0-25 cm) and from 2 to 11 for the samples from deeper layer (25-50 cm), respectively (Fig. 9B). We found no clear trend in the number of DGGE bands with increasing depth (Fig. 9B). Locality IV appears to be the richest in number of DGGE bands. We suppose that this might be due to most favorable conditions prevailing for the methanogens life as indicated by a relatively low grain median size, lower dissolved oxygen concentration or higher concentration of the ferrous iron compared to other localities (cf. Table 2).

The methanogenic community diversity in hyporheic sediment of Sitka stream was also analysed by PCR amplification, cloning and sequencing of methyl coenzyme M reductase (*mcrA*) gene. A total of 60 *mcrA* gene sequences revealed 26 different *mcrA* gene clones.

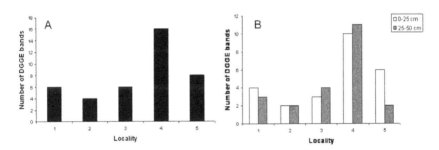

Figure 9. Number of DGGE bands associated with hyporheic sediments at two different depths along the longitudinal stream profile. A – Total number of all bands detected at each locality; B – number of bands found at different sediment depths

Most of the clones showed low affiliation with known species (< 97% nucleotide identity) and probably represented genes of novel methanogenic archeal genera/species, but all of them were closely related to uncultured methanogens from environmental samples (> 97% similarity) retrieved from BLAST. The 25 clones were clustered to four groups and were confirmed to be affiliated to *Methanosarcinales*, *Methanomicrobiales* and *Methanobacteriales* orders and other unclassified methanogens. The members of all three orders and novel methanogenic cluster were detected to occur in a whole bottom sediment irrespective of a depth, nevertheless, the richness of methanogenic archaea in the sediment was slightly higher in the upper sediment layer 0-25 cm (15 clones) than in the deeper sediment layer 25-50 cm (11 clones)(Buriánková et al. in review). The clones affiliated with *Methanomicrobiales* predominated in the deeper layer while *Methanosarcinales* clones dominated in the upper sediment layer. This prevalence of *Methanosarcinales* in the upper sediment layer was also confirmed by our FISH analyses as has been mentioned above.

4. Discussion

4.1. Occurence of methane in stream water and sediments

In spite of commonly held view of streams as well-oxygenated habitats, we found both surface and interstitial water to be supersaturated with methane compared to the atmosphere at all five localities (Mach et al. in review). Availability of interstitial habitats for bacteria and archaea carrying out anaerobic processes has been confirmed by our previous (Hlaváčová et al. 2005, 2006; Cupalová & Rulík 2007) and contemporary findings. During this study we found relatively well developed populations of methanogenic archaea at all localities and that all localities also showed positive methanogenic potential. Emissions of methane from water ecosystems results from complex microbial activity in the carbon cycle (production and consumption processes), which depends upon a large number of environmental parameters such as availability of carbon and terminal electron acceptors, flow velocity and turbulence, water depth. In our previous paper (Hlaváčová et al. 2006), we suggested that surface water concentrations, and as a consequence methane gas emissions to the atmosphere would result from downstream transport of gases by stream water (advection in/out), and moreover, from autochthonous microbial metabolism within the hyporheic zone. If so, surface water is continually saturated by gases produced by hyporheic metabolism, leading to supersaturation of surface water and induced diffusion of these gases out of river water (volatizing). Moreover, the run-off and drainage of adjacent soils can also contribute greatly to the degree of greenhouse gas supersaturation (De Angelis & Lilley 1987, Kroeze & Seitzinger 1998, Worral & Lancaster 2005, Wilcock & Sorrell 2008). For example, CH_4 in the estuarine waters may come from microbial production in water, sediment release, riverine input and inputs of methane-rich water from surrounding anoxic environments (Zhang et al. 2008b). For the European estuaries, riverine input contribute much to the estuarine CH_4 due to high CH_4 in the river waters and wetlands also play important roles. However, low CH_4 in the Changjiang Estaury (China) may be resulted from the low CH_4 in the Changjiang water together with the low net microbial production and low input from adjacent salt marshes (Zhang et al. 2008b). Dissolved methane concentrations in a surface water of Sitka stream is consistent with literature data on methane in rivers published by Middelburg et al. (2002) and Zhang et al. (2008b).

4.2. Stable carbon isotopes

A knowledge of the stable carbon isotopic ratio of methane $\delta^{13}C$-CH_4 in natural systems can be useful in studies of the mechanisms and pathways of CH_4 cycling (Sansone et al. 1997). Values of carbon isotope signature of methane ($\delta^{13}C$-CH_4) indicate biogenic nature of the methane, being usually in the range -27 ‰ up to -100 ‰ (Conrad 2004; Michener & Lajtha 2007). Whiticar et al. (1986) demonstrated that methane in freshwater sediments is isotopically distinguished by being relatively enriched in ^{13}C ($\delta^{13}C$ = -65 to -50‰) in contrast to marine sediments (-110 to -60‰). Accordingly, the two precursors of methane, namely acetate and CO_2/H_2, yield methane with markedly different $\delta^{13}C$ values; methane from acetate is relatively enriched in ^{13}C. Average minimum in the carbon isotopic composition of CII4 (61.1 ‰) occurred deeper in sediments (60 cm) while average maximum in $\delta^{13}C\text{-}CH_4$ occured in the lower sediment depth of 30 cm. Enrichment of ^{13}C in CH_4 probably reflects aerobic CH_4 oxidation because oxidation would result in residual CH_4 with $\delta^{13}C\text{-}CH_4$ values less negative than the source CH_4 (Barker & Fritz 1981; Chanton et al. 2004). However, this effect has been observed only at the study site IV.

4.3. Spatial and temporal distributions of emissions

Our working hypotesis suggested that along with the longitudinal profile of a stream, slope and flow conditions also change together with corresponding settling velocity, sediment composition and organic matter content. Thus, according to this prediction, sediment with prevalence of fine-grained particles containing higher amount of organic matter should dominate at the downstream stretches. Moreover, due to prevalence of anoxic environment, production of methane and its emissions was expected to be also higher here compared to that from upstream stretches. Based on our findings, it seems that this presumption is valid for the methane. In addition, we found higher methane concentrations in both the surface and interstitial water at the uppermost locality I compared to lower situated locality II. Similar situation with high methane concentration in the upstream part with subsequent decline further downstream was also reported from USA by Lilley et al. (1996). Dissimilarity of this first stretch is apparent in a comparison with the next, downstream laying stretch (locality II), represented by profile with steep valley and high slope. Generally, there were found very low methane concentrations either in surface or interstitial water and fluxes of emissions to atmosphere were also very low.

Flux rates of gaseous emissions into atmosphere depend on partial pressure of particular gas in the atmosphere and its concentration in a water, water temperature and further on the water depth and flow velocity. Thus, maximum peak of emissions may be expected during summer period and in well torrential stretch of the river. Silvennoinen et al. (2008), for example, found that the most upstream river site, surrounded by forests and drained peatlands, released significant amounts of CO_2 and CH_4. The downstream river sites surrounded by agricultural soils released significant amounts of N_2O whereas the CO_2 and CH_4 concentrations were low compared to the upstream site. When consider seasonal distribution of methane emissions, it is clear, in concordance with above mentioned presumption, that majority of methane emissions was relesed during a warm period of the year (81%). Effect of temperature on methane production was also observed in southeastern USA where the most methane reased to the atmosphere during warm months (Pulliam 1993). In addition, close correlation between methane emissions and temperature was reported also from south part of Baltic Sea; the temperature has been found to be a key factor driving methane emissions (Heyer & Berger 2000).

These findings also indicate that we should be very carefull in making any generalization in total emissions estimation for any given stream or river. Even though some predictions can be made based on gas concentrations measured in the surface or interstitial water, results may be very different. From this point, noteworthy was locality IV; enormous concentrations of a methane found in the deep interstitial water were caused probably by very fine, clayed sediment containing high amount of organic carbon, as well as high DOC concentrations. Supersaturation led also to the enrichment of the surface water with methane - such places may be considered as very important methane sources for surface stream and, consequently source of emissons to the atmosphere.

4.4. Benthic fluxes and potential methane oxidation

CH_4 can be produced and released into overlying near-bottom water through exchange at sediment-water interface. Methane released from the sediments into the overlying water column can be consumed by methanotrophs. Methanotrophs can oxidize as much as 100 % of methane production (Le Mer & Roger 2001). According to the season, 13-70 % of methane was consumed in a Hudson River water column (de Angelis et Scranton 1993). For the Sitka stream, measurement of benthic fluxes into the overlying surface waters indicates that methane consumption by methanotrophic bacteria is likely a dominant way of a methane loss, nevertheless some methane still supports relatively high average methane concentrations in the surface water and, in turn, high emissions to the atmosphere.

The methane production (measured as methanogenic potential) was found to be 3 orders of magnitude lower than the oxidation (methanotrophic activity), thus, almost all methane should be oxidized and consumed by methanotrophic bacteria and no methane would occur within the sediments. However, situation seems to be quite different suggesting that namely methanotrophic activity measured in a laboratory could be overestimated. Since oxidation of methane requires both available methane and oxygen, methanotrophic activity is expected to be high at sites where both methane and dissolved oxygen are available. Therefore, high values of the MA were usually found in the upper layers of the sediments (Segers 1998) or at interface between oxic and anoxic zones, respectively. Relatively high methanotrophic activity found in deeper sediments of the localities III-V indicates that methane oxidation is not restricted only to the surface sediments as is common in lakes but it also takes place at greater depths. It seems likely that oxic zone occurs in a vertical profile of the sediments and that methane diffusing from the deeper layer into the sedimentary aerobic zone is being oxidized by methanotrophs here. Increased methanotrophic activity at this hyporheic oxic-anoxic interface is probably evident also from higher abundance of type II methanotrophs in the same depth layer. Similar pathway of methane cycling has been observed by Kuivila et al. (1988) in well oxygenated sediments of Lake Washington, however, methane oxidation within the sediments would be rather normal in river sediments compared to lakes. All the above mentioned findings support our previous suggestions that coexistence of various metabolic processes in hyporheic sediments is common due to vertical and horizontal mixing of the interstitial water and occurrence of microbial biofilm (Hlaváčová et al. 2005, 2006).

4.5. Methanogens diversity

The presence of relatively rich assemblage of methanogenic archaea in hyporheic river sediments is rather surprising, however it is in accordance with other studies. The number of total different bands (i.e. estimated diversity of the methanoges) observed in the DGGE patterns of the methanogenic archaeal communities was comparable with a number of the DGGE bands found in other studies. For example, Ikenaga et al. (2004) in their study of methanogenic archaeal community in rice roots found 15-19 DGGE bands, while Watanabe et al. (2010) showed 27 bands at different positiosns in the DGGE band pattern obtained from Japanese paddy field soils. Our results from the DGGE analysis are supported by cloning and sequencing of methyl coenzyme M reductase (*mcrA*) gene which also retrieved relatively rich diversity (25 different *mcrA* gene clones) of the methanogenic community in the Sitka stream hyporheic sediments. Similar richness in number of clones was also mentioned in a methanogenic community in Zoige wetland, where 21 different clones were found (Zhang et al. 2008a), while 20 clones were described in the methane cycle of a meromictic lake in France (Biderre-Petit et al. 2011). In addition, soils from Ljubljana marsh (Slovenia) showed 17 clones (Jerman et al. 2009), for example. Both DGGE and *mcrA* gene sequencing results suggest that both hydrogenotrophic and acetoclastic methanogenesis are an integral part of the CH_4 - producing pathway in the hyporheic zone and were represented by appropriate methanogenic populations. Further, these methanogenic archaea form important component of a hyporheic microbial community and may substantially affect CH_4 cycling in the Sitka stream sediments.

5. Conclusion

To our knowledge this study is the first analysis of the composition of active methanogenic/methanotrophic communities in river hyporheic sediments. By use of various molecular methods we have shown that both methanogenic archaea and aerobic methanotrophs can be quantitatively dominant components of hyporheic biofilm community and may affect CH_4 cycling in river sediments. Their distribution within hyporheic sediments, however, only partly reflects potential methane production and consumption rates of the sediments. Rather surprising is the detection of methanotrophs in the deep sediment layer 25-50 cm, indicating that suitable conditions for methane oxidation occur here. In addition, this work constitutes the first estimation of sources, sinks and fluxes of CH_4 in the Sitka stream and in 3rd order stream environment. Fluxes of CH_4 from supersaturated interstitial sediments appear to be a main CH_4 source toward the water column. Compared with CH_4 production rates, the diffusive fluxes are very low due to efficient aerobic oxidation by methanotrophic bacteria, especially during higher flow discharges. Although fluxes to the atmosphere from the Sitka stream seems to be insignificant, they are comparable or higher in comparison with fluxes from other aquatic ecosystems, especially those measured in running waters. Finally, our results suggest that the Sitka Stream is a source of methane into the atmosphere, and loss of carbon via the fluxes of this greenhouse gas out into the ecosystem can participate significantly in river self-purification.

Author details

Martin Rulík, Adam Bednařík, Václav Mach, Lenka Brablcová, Iva Buriánková,
Pavlína Badurová and Kristýna Gratzová
*Department of Ecology and Environmental Sciences, Laboratory of Aquatic Microbial Ecology,
Faculty of Science, Palacky University in Olomouc, Czech Republic*

Acknowledgement

This work was supported by the Czech Grant Agency grant 526/09/1639 and partly by the
Ministry of Education, Youth and Sports grants 1708/G4/2009 and 2135/G4/2009 and Palacký
University IGA grant 913104161/31. We thank Lubomir Čáp and Vítězslav Maier for
analyses of dissolved methane and acetic acids, Martina Vašková and Jiří Šantrůček are
ackowledged for their help and suggestions when performing stable isotope analysis of
$^{13}C/^{12}C$ in gas samples.

6. References

Amalfitano, S. & Fazi, S., 2008: Recovery and quantification of bacterial cells associated with
    streambed sediments. – J. Microbiol. Meth. 75: 237-243.
Baker, M.A., Dahm, C.N. & Vallet, H.M., 1999: Acetate retention and metabolism in the
    hyporheic zone of a moutain stream. – Limnol. Oceanogr. 44: 1530 – 1539.
Barker, J.F. & Fritz, P., 1981: Carbon isotope fractionation during microbial methane
    oxidation. – Nature 293: 289-291.
Berner, R. A., 1980: Early Diagenesis: A Theoretical Approach. Princeton University Press.
Biderre-Petit, C. , Jézéquel, D., Dugat-Bony, E., Lopes, F., Kuever, J., Guillaume Borrel, G.,
    Viollier, E., Fonty, G. & Peyret, P., 2011: Identification of microbial communities
    involved in the methane cycle of a freshwater meromictic lake. - FEMS Microbiol. Ecol.
    77: 533–545.
Boudreau, B. P.,1996: The diffusive tortuosity of fine-grained unlithified sediments. –
    Geochim. Cosmochim. Acta 60: 3139–3142.
Bretschko, G., 1981: Vertical distribution of zoobenthos in an alpine brook of the
    RITRODAT-LUNZ study area. - Verh. Internat. Verein. Limnol. 21: 873-876.
Bretschko, G. & Klemens, W.E., 1986: Quantitative methods and aspects in the study of the
    interstitial fauna of running waters. - Stygologia 2: 297–316.
Buriánková, I. Brablcová, L., Mach, V., Hýblová, A., Badurová, P., Cupalová, J., Čáp, L. &
    Rulík, M., 2012: Methanogens and methanotrophs distribution in the
hyporheic sediments of a small lowland stream. - Fundam. Appl. Limnol. 181: 87-102.
Buriánková, I., Brablcová, L., Mach, V. & Rulík, M., in review: Methanogens diversity in
    hyporheic river sediment by targeting methyl-coenzyme M reductase (*mcrA*) gene.
Chaban, B., Ng, S.Y.M. & Jarrell, K.F., 2006: Archaeal habitats – from the extreme to the
    ordinary. - Can. J. Microbiol. 52: 73-116.

Chan, O.C., Klaus, P., Casper, P., Ulrich, A., Lueders, T. & Conrad, R., 2005: Vertical distribution of methanogenic archaeal community in Lake Dagow sediment. - Environ Microbiol 7: 1139-1149.

Carini, S., LeCleir, G. & Joye, S.B., 2005: Aerobic methane oxidation and methanotroph community composition during seasonal stratification in Mono Lake, California (USA). - Environ Microbiol 7: 1127-1138.

Chanton, J., Chaser, L., Glaser, P.& Siegel, D.I., 2001: Carbon and hydrogen isotopic effects on microbial methane from terrestrial environments. pp. 85-101. In: Flanagan, L.B., Ehleringer, J.R. and Pataki, D.E. (Eds.): Stable Isotopes and Biosphere-Atmosphere Interactions, Elsevier Press. New York.

Cicerone, R.J. & Oremland, R.S. 1988: Biogeochemical aspects of atmospheric methane. - Global Biogeochem Cy 1: 61-86.

Conrad, R., 2004: Quantification of methanogenic pathways using stable carbon isotopic signatures: a review and proposal. Organic Geochemistry 36: 739 – 752.

Conrad, R., 2007: Microbial ecology of methanogens and methanotrophs. – Adv. Agron. 96: 1-63.

Conrad, R., 2009: The global methane cycle: recent advances in understanding the microbial processes invelved. – Environ. Microbiol. Rep. 1: 285-292

Costelo, A. M., Auman, A.J., Macalady, J.L., Scow, K.M. & Lidstrom, M.E., 2002: Estimation of methanotroph abundance in a freshwater lake sediment. - Environ Microbiol 4: 443-450.

Costelo, A. M. & Lidstrom, M.E., 1999: Molecular characterization of functional and phylogenetic genes from natural populations of methanotrophs in lake sediments. – Appl. Environ. Microbiol. 65: 5066-5074.

Cupalová, J. & Rulík, M., 2007: Bacterial community analysis in river hyporheic sediments – the influence of depth and particle size. - Acta Universitatis Carolinaeana Environmentalica 21: 47-56.

de Angelis, M. A. & Lilley, M. D., 1987: Methane in surface waters of Oregon eustaries and rivers. Limnology and Oceanography 33: 716 – 722.

de Angelis, M. A. & Scranton, M. I., 1993: Fate of Methane in the Huson River and Estuary. - Global Biogeochemical Cycles 7: 509 – 523.

Duchemin, E., Lucotte, M. & Canuel, R., 1999: Comparison of static chamber and thin boundary layer equation methods for measuring greenhouse gas emissions fom large water bodies. - Environ. Sci. Technol. 33: 350-357.

Fazi, S., Amalfitano, S., Pernthaler, J. & Puddu, A., 2005: Bacterial communities associated with benthic organic matter in headwater stream microhabitats. – Environ. Microbiol. 7: 1633-1640.

Fischer, H., Kloep, F., Wilczek, S. & Pusch, M.T., 2005: A river's liver - microbial processes within the hyporheic zone of a large lowland river. - Biogeochemistry 76: 349-371.

Forster, P., Ramaswamy, V., Artaxo, P., Berntsen, T., Betts, R., Fahey D.W., Haywood, J., Lean, K., Lowe, D.C., Myhre, G., Nganga, J., Prinn, R., Raga, G., Schulz, M., Van Dorland, R., 2007: Changes in atmospheric constituents and in radiative forcing. In:

Solomon, S., Qin, D., Manning, M., Chen, Z., Marquis, M., Averyt, K.B., Tignor & M., Miller, H.L. (eds): Climate Change 2007: The Physical Science Basis. Contribution of Working Group I to the Fourth Assessment Report of the Intergovernmental Panel in Climate Change. - Cambridge University Press, Cambridge, United Kingdom/New York, NY, USA.

Ganzert, L., Jürgens, G., Münster, U. & Wagner, D., 2006: Methanogenic communities in permafrost-affected soils of the Laptev Sea coast, Siberian Arctic, characterized by 16S rRNA gene fingerprints. - FEMS Microbiol. Ecol. 59: 476-488.

Garcia, J.-L., Patel, B.K.C. & Ollivier, B., 2000: Taxonomic, phylogenetic, and ecological diversity of methanogenic Archaea. - Anaerobe 6: 205-226.

Großkopf, R., Janssen, P.H. & Liesack, W., 1998a: Diversity and structure of the methanogenic community in anoxic rice paddy soil microcosms as examined by cultivation and direct 16S rRNA gene sequence retrieval. – Appl. Environ. Microb. 64: 960–969.

Großkopf, R., Stubner, S. & Liesack, W., 1998b: Novel euryarchaeotal lineages detected on rice roots and in the anoxic soil of flooded rice microcosms. – Appl. Environ. Microb. 64: 4983–4989.

Hanson, R.S., 1998: Ecology of Methanotrophic Bacteria. - In: Burlage, R.S., Atlas, R., Stahl, D., Geesey, G. & Sayler G. (eds): Techniques in Microbial Ecology. - Oxford University Press New York, pp. 137-162.

Hanson, R.S. & Hanson, T.E., 1996: Methanotrophic bacteria. – Microbiol. Rev. 60: 439-471.

Henckel, T., Friedrich, M. & Conrad, R., 1999: Molecular analyses of the methane-oxidizing microbial community in rice field soil by targeting the genes of the 16S rRNA, particulate methane mono-oxygenase, and methanol dehydrogenase. – Appl. Environ. Microb. 65: 1980–1990.

Hendricks, S.P., 1993: Microbial ecology of the hyporheic zone: a perspective integrating hydrology and biology. - J. N. Am. Benthol. Soc. 12: 70-78.

Heyer, J. & Berger, U., 2000: Methane Emissions from the Coastal Area in the Southern Baltic Sea. - Estuarine, Coastal and Shelf Science 51: 13 – 30.

Hlaváčová, E., Rulík, M. & Čáp, L., 2005: Anaerobic microbial metabolism in hyporheic sediment of a gravel bar in a small lowland stream. – River Res. Appl. 21: 1003 – 1011.

Hlaváčová, E., Rulík, M., Čáp, L. & Mach, V., 2006: Greenhouse gases (CO_2, CH_4, N_2O) emissions to the atmosphere from a small lowland stream. – Arch. Hydrobiol. 165: 339 – 353.

Holmes, R.M., Fisher, S.G. & Grimm, N.B., 1994: Parafluvial nitrogen dynamics in a desert stream ecosystem. – J. N. Am. Benthol. Soc. 13: 468-478.

Ikenaga, M., Asakawa, S., Muraoka, Y. & Kimura, M., 2004: Methanogenic archaeal communities in rice roots grown in flooded soil pots: Estimation by PCR-DGGE and sequence analyses. - Soil. Sci. Plant Nutr. 50: 701-711.

Jeanthon, C., L'Haridon, S., Reysenbach, A.L., Corre, E., Vernet, M., Messner, P. et al., 1999: *Methanococcus vulcanius* sp. nov., a novel hyperthermophilic methanogen isolated from

East Pacific Rise, and identification of *Methanococcus* sp. DSM 4213T as *Methanococcus fervens* sp. nov. – Int. J. Syst. Evol. Microbiol. 49: 583–589.

Jerman, V., Metje, M., Mandic-Mulec, I. & Frenzel, P., 2009: Wetland restoration and methanogenesis: the activity of microbial populations and competition for substrates at different temperatures. - Biogeosciences 6: 1127-1138.

Jones, J.B. & Holmes, R.M., 1996: Surface-subsurface interactions in stream ecosystems. - TREE 11: 239 242.

Jürgens, G., Glöckner, F.-O., Amann, R., Saano, A., Montonen, L., Likolammi, M. & Münster, U. 2000: Identification of novel Archaea in bacterioplankton of a boreal forest lake by phylogenetic analysis and fluorescent in situ hybridization. - FEMS Microbiol. Ecol. 34: 45–56.

Kalyuzhnaya, M.G., Zabinsky, R., Bowerman, S., Baker, D.R., Lidstrom, M.E. & Chistoserdova, L., 2006: Fluorescence in situ hybridization-flow cytometry-cell sorting-based method for separation and enrichment of type I and type II methanotroph populations. – Appl. Environ. Microb. 72: 4293–4301.

Karl, D.M., Beversdorf, L., Björkman, K.M., Church, M.J., Martinez, A. & DeLong, E.F., 2008: Aerobic production of methane in the sea. - Nature Geoscience 1: 473-478.

Keough, B.P., Schmidt, T.M. & Hicks, R.E., 2003: Archaeal nucleic acids i n picoplankton from great lakes on three continents. – Microb. Ecol. 46: 238-248.

Kobabe, S., Wagner, D. & Pfeifer, E.M., 2004: Characterisation of microbial community composition of a Siberian tundra soil by fluorescence in situ hybridization. - FEMS Microbiol. Ecol. 50: 13-23.

Kroeze, C. & Seitzinger, S. P., 1998: The impact of land use on N_2O emissions from watersheds draining into the Northeastern Atlantic Ocean and European Seas. - Environm. Pollut. 102: 149–158.

Kuivila, K.M., Murray, J.W., Devol, A.H., Lidstrom, M.E. & Reimers, C.E., 1988: Methane cycling in the sediments of Lake Washington. – Limnol. Oceanogr. 33: 571-581.

Leichtfried, M., 1988: Bacterial substrates in gravel beds of a second order alpine stream (Project Ritrodat-Lunz, Austria). – Verh. Internat. Verein. Limnol. 23: 1325–1332.

Le Mer J. & Roger, P., 2001: Production, oxidation, emission and consumption of methane by soils. A review. - Eur. J.Soil. Biol. 37: 25-50

Lilley, M.D., de Angelis, M.A. & Olson, J.E., 1996: Methane concentrations and estimated fluxes from Pacific Northwest rivers. – Mitt. Internat. Verein. Limnol. 25: 187-196.

Lin, C., Raskin, L. & Stahl, D.A., 1997: Microbial community structure in gastrointestinal tracts of domestic animals: comparative analyzes using rRNA-targeted oligonucleotide probes. - FEMS Microbiol. Ecol. 22: 281–294.

Mach, V., Čáp, L., Buriánková, I., Brablcová, L., Cupalová, J. & Rulík, M., in review: Spatial and temporal heterogeneity in greenhouse gases emissions along longitudinal profile of a small lowland stream. – J. Limnol..

Mathrani, I.M., Boone, D.R., Mah, R.A., Fox, G.E. & Lau, P.P., 1988: *Methanohalophilus zhilinae*, sp. nov., an alkaliphilic, halophilic, methylotrophic methanogen. - Int. J. Syst. Bacteriol. 38: 139–142.

McDonald, I.R., Bodrossy, L., Chen, Y. & Murrell, C.J., 2008: Molecular ecology techniques for the study of aerobic methanotrophs. - Appl. Environ. Microbiol. 74: 1305-1315

Meyer, J.L., Likens, G.E. & Sloane, J., 1981: Phosphorus, nitrogen, and organic carbon flux in a headwater stream. – Arch. Hydrobiol. 91: 28–44.

Michener, R.& Lajtha, K., 2007: Stable Isotopes in Ecology and Environmental Science, 2nd edition. Wiley-Blackwell. ISBN: 978-1-4051-2680-9

Middelburg, J.L., Nieuwenhuize, J., Iversen, N., Høgh, N., De Wilde, H., Helder, W., Seifert, R., & Christof, O., 2002: Methane distribution in European estuaries. - Biogeochemistry 59: 95-119.

Morrice, J.A., Dahm, C.N., Valett, H.M., Unnikrishna, P. & Campana, M.E., 2000: Terminal electron accepting processes in the alluvial sediments of a headwater stream. – J. N. Am. Bethol. Soc. 19: 593-608.

Murrell, J.C., McDonald, I.R. & Bourne, D.G., 1998: Molecular methods for the study of methanotroph ecology. - FEMS Microbiol. Ecol. 27: 103-114.

Pernthaler, J., Glockne, FO, Schonhuber, W. and Amann, R. (2001). Fluorescence in situ hybridization (FISH) with rRNA-targeted oligonucleotide probes. Methods Microbiol. 30: 207-225.

Pulliam, W. M., 1993: Carbon dioxide and methane exports from a southeastern floodplain swamp. - Ecol. Monogr. 63: 29–53.

Rigby, M., Prinn, R.G., Fraser, P.J., Simmonds P.G., Langenfelds, R.L., Huang, J., Cunnold, D.M., Steele, L.P., Krummel, P.B., Weiss, R.F., O'Doherty, S., Salameh, P.K., Wang, H.J., Harth, C.M., Mühle, J., Porter, L.W., 2008: Renewed growth of atmospheric methane. - Geophysical Research Letters 35: L22805.

Rulík, M., Čáp, L. & Hlaváčová, E., 2000: Methane in the hyporheic zone of a small lowland stream (Sitka, Czech Republic). - Limnologica 30: 359 – 366.

Rulík, M. & Spáčil, R., 2004: Extracellular enzyme activity within hyporheic sediments of a small lowland stream. - Soil Biol. & Biochem. 36: 1653 – 1662.

Sanders, I.A., Heppell, C.M., Cotton, J.A., Wharton, G., Hildrew, A.G., Flowers, E.J. & Trimmer, M., 2007: Emissions of methane from chalk streams has ptential implications for agricultural practices. – Freshw. Biol 52: 1176-1186.

Sansone, F.J., Popp, B.N. & Rust, T.M., 1997: Stable carbon isotopic analysis of low-level methane in water and gas. - Anal. Chem. 69: 40-44.

Sansone, F. J., Rust, T. M. & Smith S. V., 1998: Methane distribution and cycling in Tomales bay, California. - Estuaries. 21: 66–77.

Santos Furtado, A.L. & Casper, P., 2000: Different methods for extracting bacteria from freshwater sediment and simple method to measure bacterial production in sediment samples. – J. Microbiol. Meth. 41: 249-257.

Sanz, J.L., Rodríguez, N., Díaz, E.E., Amils, R., 2011: Methanogenesis in the sediments of Rio Tinto, an extreme acidic river. - Environmental Microbiology 13: 2336-2341.

Segers, R., 1998: Methane production and methane consumption: a review of processes underlying wetland methane fluxes. - Biogeochemistry 41: 23-51.

Schindler, J.E. & Krabbenhoft, D.P., 1998: The hyporheic zone as a source of dissolved organic carbon and carbon gases to a temperate forested stream. - Biogeochemistry 43: 157-174.

Schönbauer, B. & Lewandowski, G., 1999: SEDI – a database oriented analysis and evaluation tool for processing sediment parameters. - J. Biol. Stn. Lunz 16: 13–27.

Silvennoinen, H., Liikanen, A., Rintala, J. & Martikainen, P.J., 2008: Greenhouse gas fluxes from eutrophic Temmesjoki River and its Estuary in the Liminganlahti Bay (the Baltic Sea). - Biogeochemistry 90: 193-208.

Storey, R.G., Fulthorpe, R.R., Williams, D.D., 1999: Perspectives and predictions on the microbial ecology of the hyporheic zone. - Freshwat. Biol. 41: 119-130.

Strous, M. & Jetten, M.S.M., 2004: Anaerobic oxidation of methane and ammonium. – Annu. Rev. Microbiol. 58: 99-117.

Szaran, J., 1997: Achivement of carbon isotope equilibrium in the systém HCO_3^- (solution) – CO_2 (gas). - Chemical Geology 142, 79 – 86.

Trotsenko, Y.A., Khmelenina, V.N., 2005: Aerobic methanotrophic bacteria of cold ecosystems. - FEMS Microbiol. Ecol. 53: 15-26.

Truleyová, Š., Rulík, M. & Popelka, J., 2003: Stream and interstitial water DOC of a gravel bar (Sitka stream, Czech Republic): characteristics, dynamics and presumable origin. – Arch. Hydrobiol. 158: 407–420.

Watanabe T., Asakawa, S., Nakamura, A., Nagaoka K. & Kimura, M., 2004: DGGE method for analyzing 16S rDNA of methanogenic archaeal community in paddy field soil. - FEMS Microb. Letters 232: 153-163.

Watanabe, T., Hosen, Y., Agbisit, R., Llorca, L., Fujita, D., Asakawa, S. & Kimura, M., 2010: Changes in community structure and transcriptional activity of methanogenic archaea in a paddy field soil brought about by a water-saving practice – Estimation by PCR-DGGE and qPCR of 16S rDNA and 16S rRNA. pp. 5-8. – 19th World Congress of Soil Solutions for a Changing World. 1-6 August 2010, Brisbane, Australia

Whiticar, M.J., Faber, E. & Schoell, M., 1986. Biogenic methane formation in marine and freshwater environments: CO_2 reduction vs. Acetate fermentation – isotopic evidence. - Geochemica et Cosmochimica Acta 50: 693 – 709.

Wiesenburg, D.A. & Guinasso, N.L., 1979: Equilibrium solubilities of methane, carbon monoxide, and hydrogen in water and sea water. - J. Chem. Eng. Data 24: 356–360.

Wilcock, R.J. & Sorrell, B.K., 2008: Emissions of greenhouse gases CH_4 and N_2O from low-gradient streams in agriculturally developed catchments. - Water Air Soil Poll. 188: 155-170.

Worral, F. & Lancaster, A., 2005: The release of CO_2 from riverwaters - the contribution of excess CO_2 from groundwater. - Biogeochemistry 76: 299 317.

Zhang, G., Tian, J., Jiang, N., Guo, X. Wang, Y. & Dong, X., 2008a: Methanogen community in Zoige wetland of Tibetan plateau and phenotypic characterization of a dominant uncultured methanogen cluster ZC-I. – Environ. Microbiol. 10:1850-60.

Zhang, G., Zhang, J., Liu, S., Ren, J., Xu, J., Zhang, F., 2008b: Methane in the Changjiang (Yangtze River) Estuary and its adjacent marine area: riverine input, sediments release and atmospheric fluxes. - Biogeochemistry 91: 71-84.

Chapter 9

Paenibacillus curdlanolyticus Strain B-6 Multienzyme Complex: A Novel System for Biomass Utilization

Khanok Ratanakhanokchai, Rattiya Waeonukul,
Patthra Pason, Chakrit Tachaapaikoon, Khin Lay Kyu,
Kazuo Sakka, Akihiko Kosugi and Yutaka Mori

Additional information is available at the end of the chapter

1. Introduction

To develop a bio-based economy for sustainable economic growth, it is necessary to produce chemicals and fuels from renewable resources, such as plant biomass. Plant biomass contains a complex mixture of polysaccharides, mainly cellulose and hemicellulose (mainly xylan), and other polysaccharides (Aspinall, 1980). The hemicelluloses, as well as the aromatic polymer lignin, interact with the cellulose fibrils, creating a rigid structure strengthening the plant cell wall. Therefore, complete and rapid hydrolysis of these polysaccharides requires not only cellulolytic enzymes but also the cooperation of xylanolytic enzymes (Thomson, 1993). Many microorganisms that produce enzymes capable of degrading cellulose and hemicellulose have been reported and characterized. Two enzyme systems are known for their degradation of lignocellulose by microorganisms. In many aerobic fungi and bacteria, endoglucanase, exoglucanase, and ancillary enzymes are secreted individually and can act synergistically on lignocellulose. The most thoroughly studied enzymes are the glycosyl hydrolases of *Trichoderma reesei* (Dashtban et al., 2009). On the other hand, several anaerobic cellulolytic microorganisms such as *Clostridium thermocellum* (Lamed & Bayer, 1988), *C. cellulovorans* (Doi et al., 2003), *C. josui* (Kakiuchi et al., 1998) and *C. cellulolyticum* (Gal et al., 1997) are known to produce a cell-associated, large extracellular polysaccharolytic multicomponent complex called the cellulosome, in which several cellulolytic and xylanolytic enzymes are tightly bound to a scaffolding protein (core protein). Thus, the cellulosome provides for a large variety of enzymes and attractive enzymatic properties for the degradation of recalcitrant plant biomass. So far, anaerobic microorganisms have been identified as producing the multienzyme complex, cellulosome

(Doi & Kosugi, 2004; Demain et al., 2005). However, when compared with aerobic enzymes, production of those enzymes by anaerobic culture presents a high cost because of the high price of medium, slow rate of growth and low yield of enzyme, while only a little information has been reported on cellulosome-like multienzyme complex produced by aerobic bacteria (Kim & Kim, 1993; Jiang et al,, 2004; van Dyk et al., 2009). Therefore, the multienzyme complexes, cellulosomes, produced by aerobic bacteria show great potential for improving plant biomass degradation. A facultatively anaerobic bacterium, *P. curdlanolyticus* strain B-6, is unique in that it produces extracellular xylanolytic-cellulolytic multienzyme complex under aerobic conditions (Pason et al., 2006a, 2006b; Waeonukul et al., 2009b). In the following years, the characteristics, function, genetics and mechanism of the xylanolytic-cellulolytic enzymes system of this bacterium has been the subject of considerable research. In light of new findings in this field, this review will describe the state of knowledge about the multienzyme complex of strain B-6 and its potential biotechnological exploitations.

1.1. Composition of lignocellulosic biomass

Lignocellulosic biomass is composed mainly of plant cell walls, with the structural carbohydrates, cellulose and hemicelluloses and heterogeneous phenolic polymer lignin as its primary components (Fig. 1). However, their proportions vary substantially, depending on the type, the species, and even the source of the biomass (Aspinall et al., 1980; Pérez et al., 2002; Pauly et al., 2008).

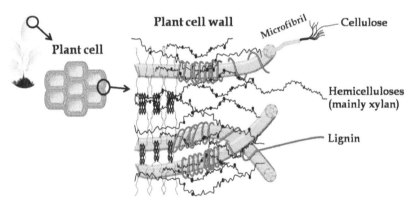

Figure 1. Structure of lignocellulosic plant biomass. (This figure is adapted from Tomme et al., 1995).

Cellulose: Cellulose, the main constituent of the plant cell wall, is a polysaccharide composed of linear glucan chains linked together by β-1,4-glycosidic bonds with cellobiose residues as the repeating unit at different degrees of polymerization, depending on resources. The cellulose chains are grouped together to form microfibrils, which are bundled together to form cellulose fibers. The cellulose microfibrils are mostly independent but the ultrastructure of cellulose is largely due to the presence of covalent bonds, hydrogen bonds and Van der Waals forces. Hydrogen bonding within a cellulose microfibril determines 'straightness' of the chain but inter-chain hydrogen bonds might introduce order (crystalline) or disorder (amorphous) into the structure of the cellulose (Klemm et al., 2005). In the latter conformation, cellulose is more susceptible to enzymatic degradation (Pérez et al., 2002). In nature, cellulose appears to be associated with other plant compounds and this association may affect its biodegradation.

Hemicelluloses: Hemicelluloses are the second most abundant polymers and differ from cellulose in that they are not chemically homogeneous. Hemicelluloses are branched, heterogenous polymers of pentoses (xylose, arabinose), hexoses (mannose, glucose, galactose) and acetylated sugars. They have lower molecular weight compared to cellulose and branches with short lateral chains that are easily hydrolysed (Saha, 2003; Scheller & Ulvskov, 2010). Hemicelluloses differ in composition. Hemicelluloses in agricultural biomass like straws and grasses are composed mainly of xylan, while softwood hemicelluloses contain mainly glucomannan. In many plants, xylans are heteropolysaccharides with backbone chains of 1,4-linked β-D-xylopyranose units. In addition to xylose, xylan may contain arabinose, glucuronic acid, or its 4-O-methyl ether, acetic acid, ferulic and p-coumaric acids. Hemicelluloses are bound via hydrogen bonds to the cellulose microfibrils in the plant cell wall, crosslinking them into a robust network. Hemicelluloses are also covalently attached to lignin, forming together with cellulose to form a highly complex structure.

Lignin: Lignin is the third most abundant polymer in nature. It is present in plant cell walls and confers a rigid, impermeable, resistance to microbial attack and oxidative stress. Lignin is a complex network formed by polymerization of phenyl propane units and constitutes the most abundant non-polysaccharide fraction in lignocelluloses (Pérez et al., 2002; Sánchez, 2009). The three monomers in lignin are p-coumaryl alcohol, coniferyl alcohol and sinapyl alcohol; they are joined through alkyl–aryl, alkyl–alkyl and aryl–aryl ether bonds. Lignin embeds the cellulose thereby offering protection against microbial and enzymatic degradation. Furthermore, lignin is able to form covalent bonds to some hemicelluloses, e.g. benzyl ester bonds with the carboxyl group of 4-O-methyl-D-glucuronic acid in xylan. More stable ether bonds, also known as lignin carbohydrate complexes, can be formed between lignin and arabinose, or between galactose side groups in xylans and mannans.

1.2. Biodegradation of lignocellulosic biomass

Several biological methods for lignocellulose recycling based on the enzymology of cellulose, hemicelluloses, and lignin degradation have been developed. To date, processes that use lignocellulolytic enzymes or microorganisms could lead to promising, environmentally friendly technologies. The relationship between cellulose and hemicellulose in the cell walls of higher plants is much more intimate than was previously thought. It is possible that molecules at the cellulose-hemicellulose boundaries, and those within the crystalline cellulose, require different enzymes for efficient hydrolysis.

Cellulase: Cellulases responsible for the hydrolysis of cellulose are composed of a complex mixture of enzymes with different specificities to hydrolyze the β-1,4-glycosidic linkages (Fig. 2A). Cellulases can be divided into three major enzyme activity classes (Goyal et al., 1991; Rabinovich et al., 2002). These are endoglucanases or endo-1-4-β-glucanase (EC 3.2.1.4), exoglucanase or cellobiohydrolase (EC 3.2.91), and β-glucosidase (EC 3.2.1.21). Endoglucanases, are thought to initiate attack randomly at multiple internal sites in the amorphous regions of the cellulose fiber, which opens-up sites for subsequent attack by the cellobiohydrolases. Cellobiohydrolases remove cellobiose from the ends of both sides of the glucan chain. Moreover, cellobiohydrolase can hydrolyze highly crystalline cellulose. β-glucosidase hydrolyzes cellobiose and in some cases short chain cellooligosaccharides to glucose.

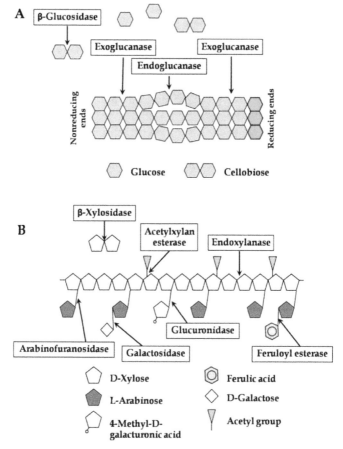

Figure 2. Enzyme systems involved in the degradation of cellulose (A) and xylan (B). (This figure is adapted from Aro et al., 2005).

Hemicellulase: Xylan is the main carbohydrate found in hemicelluloses. Its complete degradation requires the cooperative action of a variety of hydrolytic enzymes (Fig. 2B). Xylanases are frequently classified according to their action on distinct substrates: endo-1,4-β-xylanase (endoxylanase) (EC 3.2.1.8) generates xylooligosaccharides from the cleavage of xylan while 1,4-β-xylosidase (EC 3.2.1.37) produces xylose from xylobiose and short chain xylooligosaccharides. In addition, xylan degradation needs accessory enzymes, such as α-L-arabinofuranosidase (EC 3.2.1.55), α 4 O methyl D glucuronidase (EC 3.2.1.39), acetyl xylan esterase (EC 3.1.1.72), ferulic acid esterase (EC 3.1.1.73), and p-coumaric acid esterase (EC 3.1.1.-), acting synergistically, to efficiently hydrolyze wood xylans. In the case of acetyl-4-O-methylglucuronoxylan, which is one of the most common hemicelluloses, four different enzymes are required for degradation: endo-1,4-β-xylanase, acetyl esterase (EC 3.1.1.6), α-glucuronidase, and β-xylosidase. The degradation of O-acetyl galactoglucomannan starts with the rupture of the polymer by endomannanase (EC 3.2.1.78). Acetylglucomannan esterase (EC 3.1.1.-) removes acetyl groups, and α-galactosidase (EC 3.2.1.22) eliminates galactose residues. Finally, β-mannosidase (EC 3.2.1.85) and β-glucosidase break down the endomannanase-generated oligomeric β-1,4 bonds (Thomson, 1993; Li et al., 2000; Pérez et al., 2002).

1.3. Multienzyme complex cellulosome

The enzyme systems for the lignocellulose degradation by microorganisms can be generally regarded as non-complexed or complexed enzymes (Lynd et al., 2002). In the case of aerobic fungi and bacteria, the cellulase enzymes are free and mostly secreted. In such organisms, by the very nature of the growth of the organisms, they are able to reach and penetrate the cellulosic substrate and, hence, the secreted cellulases are capable of hydrolyzing the substrate. The enzymes in these cases are not organized into high molecular weight complexes and are called non-complexed (Fig. 3A). The polysaccharide hydrolases of the aerobic fungi are largely described based on the examples from *Trichoderma, Penicillum, Fusarium, Humicola, Phanerochaete,* etc., where a large number of the cellulases are encountered (Dashtban et al., 2009; Sánchez, 2009). In contrast, various cellulases and hemicellulases from several anaerobic cellulolytic microorganisms, are tightly bound to a scaffolding protein, as core protein and organized to form structures on the cell surfaces; these systems are called complexed enzymes or cellulosomes (Fig. 3B). The cellulosome is thought to allow concerted enzyme activities in close proximity to the bacterial cell, enabling optimum synergism between the enzymes presented on the cellulosome. Concomitantly, the cellulosome also minimizes the distance over which hydrolysis products must diffuse, allowing efficient uptake of these oligosaccharides by the host cells (Bayer et al., 1994; Schwarz, 2001; Lynd et al., 2002).

Biotechnological applications in terms of hydrolysis efficiency for complexed enzyme systems might have an advantage over non-complexed enzyme systems. The high efficiency of the cellulosome has been attributed to (i) the correct ratio between catalytic domains that optimize synergism between them, (ii) appropriate spacing between the individual components to further favor synergism, (iii) the presence of different enzymatic activities (cellulolytic or hemicellulolytic enzymes) in the cellulosome that can remove "physical hindrances" of other polysaccharides in heterogeneous plant cell materials (Lynd et al., 2002), and (iv) the presence of carbohydrate-binding modules (CBMs) that can increase the rate of hydrolysis by bringing the cellulosome into intimate and prolonged association with its recalcitrant substrate (Shoseyov et al., 2006). Thus, the complexed enzyme system, cellulosome, may provide great potential for the degradation of plant biomass.

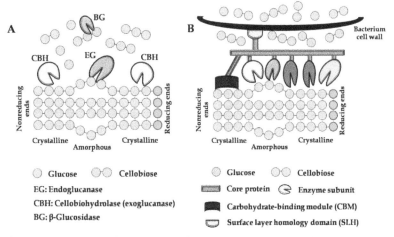

Figure 3. Simplified schematic of the hydrolysis of amorphous and microcrystalline celluloses by non-complexed (A) and complexed (B) cellulase systems. (This figure is adapted from Lynd et al., 2002).

The cellulosome was first identified in 1983 from the anaerobic, thermophilic, spore-forming *Clostridium thermocellum* (Lamed et al., 1983). The cellulosome of *C. thermocellum* is commonly studied along with cellulosomes from the anaerobic mesophiles, *C. cellulovorans* (Doi et al., 2003), *C. josui* (Kakiuchi et al., 1998) and *C. cellulolyticum* (Gal et al., 1997). All cellulosomes share similar characteristics, they all contain a large distinct protein, referred to as the scaffoldin which allows binding of the whole complex to microcrystalline cellulose via CBM. Also, the cellulosome scaffoldin expresses type I cohesins which allow binding of a wide variety of cellulolytic and hemicellulolytic enzymes within the complex via the expression of complementary type I dockerins on enzymes. Similarly, at the C-terminal the scaffoldin expresses type II cohesins, which allow the binding of the cellulosome to the cell through type II dockerins on surface layer homology proteins (SLH) (Fig. 4).

Cellulosomes are produced mainly by anaerobic bacteria, mostly from the class clostridia, and some anaerobic fungi such as genus *Neocallimastix* (Dalrymple et al., 1997), *Piromyces* (Teunissen et al., 1991) and *Orpinomyces* (Li et al., 1997). However, evidence suggests the presence of cellulosomes or cellulosome-like multienzyme complexes in a few aerobic microorganisms (Table 1). It is speculated that several other cellulolytic bacteria may also produce cellulosomes not yet described.

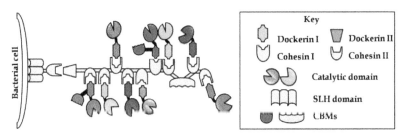

Figure 4. Simplified schematic of general cellulosome components and connection with cell surface based on knowledge of *Clostridium* cellulosome. (This figure is adapted from Bayer et al., 1994).

Anaerobic			Aerobic		
Microorganism	Source	Ref.	Microorganism	Source	Ref.
Bacteria			*Bacteria*		
Acetivibrio cellulolyticus	Sewage	Ding et al., 1999	*Bacillus circulans* F-2	Potato starch granules	Kim and Kim, 1993
Amorocellulobacter alkalithermophilum	Soil	Watthanalarm-lort et al., 2012	*Bacillus licheniformis* SVD1	Bioreactor	van Dyk et al., 2009
Bacteroides cellulosolvens	Sewage	Ding et al., 2000	*Paenibacillus curdlanolyticus* B-6	Anaerobic digester	Pason et al., 2006b
Bacteroides sp. strain P-1	Anaerobic digester	Ponpium et al., 2000	*Sorangium cellulosum*	Soil	Hou, et al., 2006
Butyrivibrio fibrisolvens	Rumen	Berger et al., 1990			
Clostridium acetobutylicum	Soil	Sabathé et al.,2002			
Clostridium cellobioparum	Rumen	Lamed et al., 1987	*Actinomycetes*		
Clostridium cellulolyticum	Compost	Pagès et al., 1997	*Streptomyces olivaceoviridis* E-86	Soil	Jiang et al., 2004
Clostridium cellulovorans	Fermenter	Sleat et al., 1984			
Clostridium josui	Compost	Kakiuchi et al., 1998	*Fungi*		
Clostridium papyrosolvens	Paper mill	Pohlschröder et al., 1994	*Chaetomium* sp. Nov. MS-017	Rotted wood	Ohtsuki et al., 2005
Clostridium thermocellum	Sewage soil	Lamed et al., 1983			
Eubacterium cellulosolvens	Rumen	Blair and Anderson, 1999b			
Ruminococcus albus	Rumen	Ohara et al., 2000			
Ruminococcus flavefaciens	Rumen	Ding et al., 2001			
Tepidimicrobium xylanilyticum BT14	Soil	Phitsuwan et al., 2010			

Anaerobic			Aerobic		
Microorganism	Source	Ref.	Microorganism	Source	Ref.
Thermoanaerobacterium thermosaccharolyticum NOI-1	Soil	Chimtong et al., 2011			
Fungi					
Neocallimastix patriciarum	Rumen	Dalrymple et al., 1997			
Orpinomyces joyonii	Rumen	Qiu et al., 2000			
Orpinomyces PC-2	Rumen	Borneman et al., 1989			
Piromyces equi	Rumen	Teunissen et al., 1991			
Piromyces E2	Faeces	Teunissen et al., 1991			

Table 1. Cellulosome and cellulosome-like multienzyme complexes from anaerobic and aerobic microorganisms. (This table is adapted from Doi & Kosugi, 2004).

2. Novel multienzyme complex system from *P. curdlanolyticus* strain B-6

Efficient enzymatic degradation of lignocellulosic biomass requires a tight interaction between the enzymes and their substrates, and the cooperation of multiple enzymes to enhance the hydrolysis due to the complex structure. Multienzyme complexes, cellulosomes from anaerobic cellulolytic microorganisms, are dedicated to hydrolyzing lignocellulosic substances efficiently because of a large variety of cellulases and hemicellulases in complexes, useful enzymatic properties, and binding ability to insoluble cellulose and/or xylan via CBMs (Bayer et al., 2004; Doi and Kosugi, 2004; Schwarz et al., 2001; Shoham et al., 1999). When compared with aerobic enzymes, they produce several individual enzymes, but microorganisms are not binding to insoluble substrates. However, *P. curdlanolyticus* B-6 was found to produce a multienzyme complex under aerobic conditions (Pason et al., 2006a, 2006b). Little information has been reported on cellulosome-like multienzyme complexes produced by aerobic bacterium (Kim & Kim, 1993; Jiang et al., 2004; van Dyk et al., 2009). Therefore, the multienzyme complex produced by strain B-6 is critical for improving plant biomass degradation.

2.1. Selection of multienzyme complex-producing bacteria under aerobic cultivation

Among several *Bacillus* strains, isolated from various sources and cultivated under aerobic conditions, *P. curdlanolyticus* strain B-6 shows important evidences for multienzyme complex producing bacterium (Pason et al., 2006a) as follows: high production of cellulase and xylanase, presence of CBMs that have ability to bind to insoluble substances, adhesion of bacterial cells to insoluble substances, and production of multiple cellulases and xylanases in the form of a high molecular weight complex. Thus, strain B-6 exhibits great promise bacterium in the production of multienzyme complex under aerobic conditions. Some properties of bacterial cells and cellulase and xylanase from strain B-6 compared with other *Bacillus* spp. are shown in Table 2.

Strain (*Bacillus* sp.) and growth condition	Specific activity (U/mg protein)		Enzyme binding ability to insoluble substances (%)		Adhesion of cells to insoluble substances (%)		Zymogram analysis	
	CMCase	Xylanase	Avicel	Xylan	Avicel	Xylan	CMCase band	Xylanase band
1. Strain B-6								
Avicel grown	0.16	1.12	57.1	64.3	28.0	39.9	11	13
Xylan grown	0.12	7.19	39.1	51.5	13.6	74.7	9	15
2. Strain H-4								
Avicel grown	0.15	1.10	50.0	50.0	0	0	2	2
Xylan grown	0.09	4.23	31.1	38.5	0	0	1	3
3. Strain S-1								
Avicel grown	0.15	0.90	43.4	49.1	0	0	3	2
Xylan grown	0.09	4.49	37.9	45.8	0	0	2	3
4. Strain X-11								
Avicel grown	0	0	0	0	0	0	0	0
Xylan grown	0.05	3.29	29.2	45.0	0	0	0	2
5. Strain X-24								
Avicel grown	0	0	0	0	0	0	0	0
Xylan grown	0.06	3.19	29.6	36.1	0	0	0	2
6. Stain X-26								
Avicel grown	0	0	0	0	0	0	0	0
Xylan grown	0.04	3.10	28.2	38.2	0	0	0	2

Table 2. Production of carboxymethyl cellulase (CMCase) and xylanase by *Bacillus* strains; binding ability of enzymes to insoluble substances; adherence of bacterial cells to insoluble substances; and zymograms analysis in culture supernatant.

P. curdlanolyticus strain B-6 was a facultative, spore-forming, Gram-positive, motile, rod-shaped organism and produced catalase. Thus, this bacterium was identified as a member of the genus *Bacillus* according to Bergey's Manual of Systematic Bacteriology (Sneath, 1986). The bacterium was also identified by 16S rRNA gene sequence analysis. The use of a specific PCR primer designed for differentiating the genus *Paenibacillus* from other members of the *Bacillaceae* showed that this strain had the same amplified 16S rRNA gene fragment as a member of the genus *Paenibacillus*. Based on these observations, it is reckoned that this strain was transferred to the genus *Paenibacillus* (Shida et al., 1997). The 16S rDNA sequence of this strain had 1,424 base pairs and 97% similarity with *Paenibacillus curdlanolyticus* (Innis & Gelfand, 1990).

2.2. Characteristics of *P. curdlanolyticus* B-6 multienzyme complex

During growth of *P. curdlanolyticus* B-6 on Berg's mineral salt medium containing 0.5% xylan as carbon sources, the protein concentration in the medium was low up to the late stationary growth phase. CMCase and xylanase activities could be detected in the culture medium after the late exponential phase (Pason et al., 2006b). At the declining growth phase, the extracellular xylanase and CMCase rapidly increased due to the release of enzymes from the cell surfaces into the culture medium. These phenomena were different from the growth patterns of other aerobic bacteria, which grew and produced extracellular enzymes into culture supernatant immediately, but similar to those of the anaerobic bacteria which produced multienzyme complexes (cellulosomes) around the cell surfaces and adhered to these substrates and secreted into culture supernatant later (Bayer & Lamed, 1986; Lamed & Bayer, 1988). The observation of cell surfaces at the late exponential growth phase by scanning electron microscopy (SEM) revealed that the cells adhered to xylan (Fig. 5A), similar to the cells of the cellulosome producing anaerobic bacterium, *C. thermocellum*, which is a cell associated entity that mediates the adhesion of the bacterium to cellulose (Lamed et al., 1987; Mayer et al., 1987), whereas the surface of the cells of strain B-6 at the late stationary growth phase lacked such structures because the multienzyme complex was released into the medium from the cell surfaces (Fig. 5B). In addition, the pattern of multienzyme complex in the culture medium at the late stationary growth phase was determined. Native-polyacrylamide gel electrophoresis (native-PAGE) exhibited a high molecular weight band at the top of the gel (Fig. 6, lane 1). This protein band was dissociated into major and minor components through treatment by boiling in sodium dodecyl sulphate (SDS) solution, showing at least 18 proteins with molecular masses in the range of 29 to 280 kDa (Fig. 6, lane 2). Among those protein bands, at least 15 bands showed xylanase activities (Fig. 6, lane 3) and at least 9 bands showed CMCase activities (Fig. 6, lane 4) on zymograms. These multiple cellulases and xylanases are assembled into the high molecular weight complexes and released from the cell surfaces into medium at the late stationary growth phase. In *C. thermocellum*, the cellulosome consisted of many different types of glycosyl hydrolases, including cellulases, hemicellulases, and carbohydrate esterases, which served to promote their synergistic action (Lamed et al., 1983). These evidences confirm that the strain B-6 can produce xylanolytic-cellulolytic enzyme system that exists as multienzyme complex under aerobic conditions.

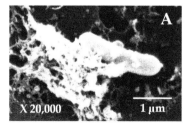 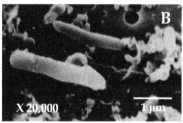

Figure 5. SEM of the cell surfaces of *P. curdlanolyticus* B-6 harvested at the late exponential growth phase showing adhesion of cell to xylan (A) and the cell harvested at the late stationary growth phase showing no adhesion of cell to xylan (B).

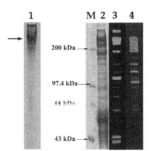

Figure 6. Proteins and enzymes patterns of multienzyme complex in culture supernatant at the late stationary growth phase; Native-PAGE (lane 1), SDS-PAGE (lane 2), and zymograms analysis of xylanase activity (lane 3), and CMCase activity (lane 4).

2.3. Effect of carbon sources on the induction of multienzyme complex in *P. curdlanolyticus* B-6

The effect of polymeric substances such as cellulose, xylan, corn hull, and sugarcane bagasse, and of soluble sugars such as L-arabinose, D-galactose, D-glucose, D-xylose, and cellobiose on the induction of multienzyme complexes in a facultatively anaerobic bacterium, *P. curdlanolyticus* B-6, was investigated under aerobic conditions (Waeonukul et al., 2008; 2009b). Cells grown on each carbon source adhered to cellulose. Hence strain B-6 cells from all carbon sources must have an essential component responsible for anchoring the cells to the substrate surfaces. Native–PAGE, SDS–PAGE, zymograms analysis, and enzymatic assays revealed that many proteins having xylanolytic and cellulolytic activities from *P. curdlanolyticus* B-6 grown on each carbon source were produced as multienzyme complex into the culture supernatants. These results indicated that strain B-6 produced multienzyme complexes when grown on both polymeric substances and soluble sugars. However, the subunits expressed in the multienzyme complex of strain B-6 depended on the carbon sources. These observations are consistent with previous reports that the enzymatic activities and enzyme compositions of the cellulosomes of *C. thermocellum* (Bayer et al., 1985; Bhat et al., 1993; Nochur et al., 1993), *C. cellulolyticum* (Mohand-Oussaid et al., 1999), and *C. cellulovorans* (Kosugi et al., 2001; Han et al., 2004; 2005) and the xylanosome of *S. olivaceoviridis* E-86 (Jiang et al., 2004) were affected by carbon sources in the media.

Many investigators have reported that the synthesis of cellulosome assemblies requires the presence of crystalline cellulose under anaerobic conditions, and that synthesis hardly occurs in growth on glucose or other soluble carbohydrates (Nochur et al., 1992; Blair & Anderson; 1999a; Bayer 2004; Doi & Kosugi, 2004). Some strains of *C. thermocellum* (Bayer et al., 1985; Bhat et al., 1993), however, can induce cellulosome synthesis when grown on cellobiose. *P. curdlanolyticus* B-6 differs from most cellulosome-producing microorganisms in that it produces multienzyme complex when grown on both polymeric substances and soluble sugars under aerobic conditions. Therefore, the mechanism of multienzyme complex formation by strain B-6 must be different from that of other microorganisms.

3. The feature of *P. curdlanolyticus* B-6 multienzyme complex

Recently, the structures and mechanisms for assembly of multienzyme complexes, cellulosomes, in anaerobic cellulolytic microorganisms are clear (Bayer et al., 2004, 2007; Doi & Kosugi, 2004). Generally, the key feature of the cellulosome is a scaffoldin that integrates the various catalytic subunits into the complex by self-assembly by cohesion-dockerin interaction. However, the structure and mechanism of the multienzyme complex produced by a facultatively anaerobic bacterium, such as *P. curdlanolyticus* B-6 is still unknown. In order to describe features of the multienzyme complex system produced by strain B-6, the multienzyme complex was purified by four kinds of chromatography (cellulose affinity, gel filtration, anion-exchange and hydrophobic-interaction chromatographys) (Fig. 7).

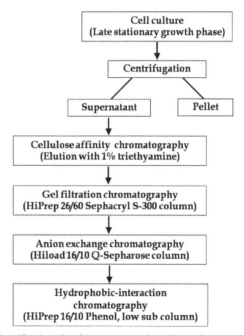

Figure 7. Isolation and purification of multienzyme complex of *P. curdlanolyticus* strain B-6.

The multienzyme complex of *P. curdlanolyticus* strain B-6 with molecular mass of 1,450 (G1) was isolated from culture supernatant at the late stationary growth phase through cellulose affinity and Sephacryl S-300 gel filtration chromatographys (Pason et al., 2006b). Basically, the individual cellulosomes from anaerobic bacteria show 600 kDa to 2.1 MDa complexes size and show cohesion-dockerin domain as a signature protein (Bayer et al., 2004; Doi & Kosugi, 2004). While, multienzyme complexes from aerobic microorganisms, were range in mass from about 468 kDa to 2 MDa (with contained 5-12 protein subunits) (Table 3) and has no report of cohesion-dockerin domain. Here, the multienzyme complex produced by strain B-6 under aerobic conditions was the first report on characterization.

Multienzyme complex	Mol. Mass (kDa)	Protein subunits	Ref.
Aerobic microorganisms			
Paenibacillus curdlanolyticus B-6	1450	11	Pason et al.,2006b
Bacillus circulans F-2	669	7	Kim and Kim, 1993
Bacillus licheniformis SVD1	2000	12	van Dyk et al., 2009
Sorangium cellulosum	1000-2000	10	Hou et al., 2006
Streptomyces olivaceoviridis E-86	1200	5	Jiang et al., 2004
Chaetomium sp. Nov. MS-017	468	12	Ohtsuki et al., 2005
Anaerobic microorganisms			
Clostridium acetobutylicum	665	11	Sabathé et al.,2002
Clostridium cellulolyticum	600	14	Gal et al., 1997
Clostridium cellulovorans	900	10	Shoseyov & Doi 1990
Clostridium josui	700	14	Kakiuchi et al., 1998
Clostridium popyrosolvens	600	15	Pohlschröder et al., 1994
Clostridium thermocellum	2100	14	Lamed et al., 1983
Ruminococcus albus	1500	15	Ohara et al., 2000

Table 3. Molecular weights and protein subunits of multienzyme complexes from aerobic and anaerobic microorganisms.

Elucidation of the purified multienzyme feature of *P. curdlanolyticus* strain B-6 was followed by anion-exchange and hydrophobic-interaction chromatographys (Pason et al., 2010). The complex G1 from gel filtration chromatography (1,450 kDa) was purified by anion-exchange chromatography and showed at least five large protein complexes or aggregates, namely F1-F5. Among the fractions obtained from anion-exchange chromatography, F1 was apparently the most suited fraction to study on the organization and function of the multienzyme system of strain B-6 because F1 formed one clear band on the top of native PAGE, had the highest xylanase activity, and its subunit composition was clearly shown on SDS-PAGE. In the final step, complex F1 was separated to one major complex (H1) and two minor protein components (H2 and H3) by hydrophobic-interaction chromatography. The multienzyme complex (H1) was composed of a 280 kDa protein with xylanase activity, a 260 kDa protein that is a truncated form on the C-terminal side of the 280 kDa protein, two xylanases of 40 and 48 kDa, and 60 and 65 kDa proteins having both xylanase and CMCase activities (Fig. 8). The two components (280 and 40 kDa) of the multienzyme complex has characteristics similar to the cellulosome of *C. thermocellum* in that it is composed of a scaffolding protein and a catalytic subunit (Bayer et al., 1998; Demain et al., 2005). The 280 kDa protein resembled the scaffolding proteins of the multienzyme complex based on its migratory behavior in polyacrylamide gels and as a glycoprotein. The 280 kDa protein and a 40 kDa major xylanase subunit are the key components of multienzyme complex of the strain B-6.

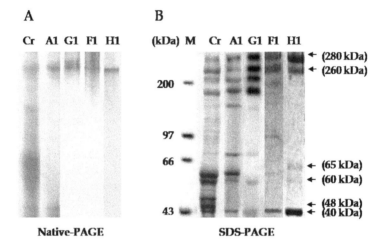

Figure 8. Native-PAGE (A) and SDS-PAGE (B) in isolated complex from culture supernatant at the late stationary growth phase (lane Cr), affinity column (lane A1), gel filtration column (lane G1), anion-exchange column (lane F1) and hydrophobic-interaction column (lane H1). All samples contained 200 μg of protein.

These apparently propose that *P. curdlanolyticus* B-6 produced multienzyme complex, which consisted of many subunit compositions. The large protein (280 kDa) may function as a scaffoldin-like protein that allowed the enzyme subunits, majority is 40 kDa, binding to form a multienzyme complex. The key components, 280 and 40 kDa, are identified in the next topic.

4. Molecular structure of important xylanases

P. curdlanolyticus B-6 produces an extracellular xylanolytic-cellulolytic multienzyme complex mainly comprised of xylanases under aerobic conditions. To understand the xylanase system, a genomic library of the strain B-6 was constructed and screened for high xylanase activity. Recently, six xylanase genes, *S1* (Pason et al., 2010), *xyn10A* (Waeonukul et al., 2009a), *xyn10B* (Sudo et al., 2010), *xyn10C* (unpublished data), *xyn10D* (Sakka et al., 2011) and *xyn11A* (Pason et al., 2010) were cloned, and the translated products were characterized (Table 4).

Enzyme	Modular structure	GH family	Mol. Mass (kDa)	GenBank accession No.
S1	SLH SLH No homology sequence	U	91	-
Xyn10A	CBM22 CBM22 CBM22 GH10 CBM9 SLH SLH SLH	10	142	EU418764
Xyn10B	GH10	10	40	AB570291
Xyn10C	GH10	10	35	AB688987
Xyn10D	GH10 Fn3 CBM3	10	61	AB600191
Xyn11A	GH11 CBM36	11	40	FJ956758

Abbreviations: CBM, carbohydrate-binding module; Fn, fibronectin homology module; GH, glycosyl hydrolase; SLH, surface layer homology domain; U, unhomology sequence

Table 4. Modular structure xylanases of *P. curdlanolyticus* strain B-6.

S1 protein: From the early research, the 280 kDa subunit (S1) plays a role of scaffoldin in assembling the enzyme complex and shows xylanase activity (Pason et al., 2010). The *S1* gene consists of 2,589 nucleotides and encodes 863 amino acids with a molecular weight of 91,000 Da, indicating that the 280 kDa subunit is highly glycosylated. Sequence analysis revealed that S1 did not have significant homology with any proteins in the databases except for two surface layer homology (SLH) domains in its N-terminal region. Surprisingly, the recombinant S1 exhibits xylanase activity, and cellulose- and xylan-binding ability, suggesting that the S1 should be a novel xylanase and CBM(s) with new functions (unpublished data).

Xylanase Xyn10A: The *xyn10A* gene consists of 3,828 nucleotides encoding a protein of 1,276 amino acids with a predicted molecular weight of 142,726 Da. Xyn10A is a multidomain enzyme comprised of nine domains in the following order: three family-22 CBMs, a family-10 catalytic domain of glycosyl hydrolases (GH), a family-9 CBM, a glycine-rich region, and three SLH domains. Xyn10A can effectively hydrolyze insoluble xylan and natural biomass without pretreatment such as sugarcane bagasse, corn hull, rice bran, rice husk and rice straw. Xyn10A binds to various insoluble polysaccharides such as cellulose, xylan and chitin. The SLH domains functioned in Xyn10A by anchoring this enzyme to the cell surfaces of *P. curdlanolyticus* B-6. Removal of the CBMs from Xyn10A strongly reduced the ability of binding and plant cell wall hydrolysis. Therefore, the CBMs of Xyn10A play an important role in the hydrolysis of native biomass materials (Waeonukul et al., 2009a).

Xylanase Xyn10B: The *xyn10B* gene consists of 1,047 nucleotides encoding a protein of 349 amino acids with a predicted molecular weight of 40,480 Da. Xyn10B consists of only a family-10 catalytic of GH. Xyn10B is an intracellular endoxylanase (Sudo et al., 2010).

Xylanase Xyn10C: The *xyn10C* gene consists of 957 nucleotides and encodes 318 amino acid residues with a predicted molecular weight of 35,123 Da. Xyn10C is a single module enzyme consisting of a signal peptide and a family-10 catalytic module of GH (unpublished data).

Xylanase Xyn10D: The *xyn10D* gene consists of 1,734 nucleotides and encodes 577 amino acid residues with a calculated molecular weight of 61,811 Da. Xylanase Xyn10D is a modular enzyme consisting of a family-10 catalytic module of the GH, a fibronectin type-3 homology (Fn3) module, and family-3 CBM, in that order, from the N terminus. The CBM3 in Xyn10D has an affinity for cellulose and xylan, and plays an important role in hydrolysis of arabinoxylan and native biomass materials (Sakka et al., 2011).

Xylanase Xyn11A: The *xyn11A* gene consists of 1,150 bp and encodes a protein of 385 amino acids with a molecular weight of 40,000 Da. Xyn11A is composed of two major functional domains, a catalytic domain belonging to family-11 GH and a CBM classified as family-36. A glycine- and asparagine-repeated sequences existed between the two domains. Xyn11A has been identified to be one of the major xylanase subunit in the multienzyme complex of strain B-6 (Pason et al., 2010).

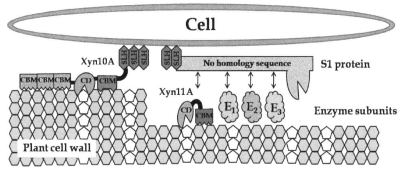

Figure 9. Simplified schematic view of the interaction between the *P. curdlanolyticus* B-6 multienzyme complex system and its substrate, and its connection to the cell surface via an associated anchoring protein. (Abbreviations: CBM, carbohydrate-binding module; CD, catalytic domain, E$_n$, enzyme subunit; SLH, surface layer homology domain).

Based on both biochemical and molecular biological findings, a simplistic schematic view of the enzyme system from *P. curdlanolyticus* B-6 and its interaction with substrate and cell surface was created and presented in Fig. 9. In this assessment, the S1 protein did not have significant homology with any proteins in the databases except for two S-layer homology domains in its N-terminal region. However, the S1 protein that exhibits xylanase activity and cellulose- and xylan-binding ability, and contains cell anchoring function, seems remarkable. The multifunctional protein S1 is also responsible for forming the enzyme subunits into the complex and anchoring the complex into cell surface via the SLH domain. The interaction between S1 protein and enzyme subunits should be a mechanism distinct from the cohesion-dockerin interaction known in cellulosome of anaerobic microorganisms, since cohesion- or dockerin– like sequences were not observed in the S1 protein or the major xylanase subunit, Xyn11A. In addition, strain B-6 also produces cell bound multimodular xylanase Xyn10A that contains the numerous CBMs and SLH domains. Xyn10A can bind to the plant cell wall through CBM, whereas the catalytic module (GH10) is able to access its target substrate. Thus, the CBM greatly increases the concentration of the enzyme in the vicinity of the substrate, leading to the observed increase in polysaccharide hydrolysis. Besides, the presence of the functional CBMs and SLH domains in Xyn10A allows the cells to attach to substrate. Although, the overall structure of the enzyme complex system of the strain B-6 is not entirely clear, the enzyme complex has unique characteristics distinct from multienzyme complex cellulosome of anaerobic microorganisms. However, the mechanism for complex formation, interaction between the S1 protein as scaffoldin and enzyme subunits, needs to be further investigated.

5. Biotechnological uses of *P. curdlanolyticus* B-6 multienzyme complex

Biological conversion of lignocellulosic materials has been proposed as a renewable and sustainable route for the production of value-added products (Bayer et al., 2007, Doi et al., 2003). There is much interest in exploiting the properties of multienzyme complexes for practical purposes. The facultative bacterium, *P. curdlanolyticus* strain B-6 produces a unique extracellular multienzyme system under aerobic conditions that effectively degrade cellulose and hemicellulose by gaining access through the protective matrix surrounding the cellulose microfibrils of plant cell walls. Therefore, the multienzyme complex from strain B-6 is a promising enzyme which can potentially be used in many applications, such as enhancing extraction and production of value-added bioproducts by saccharification of cell wall components and application for construction of the modular enzymes creation (Fig. 10).

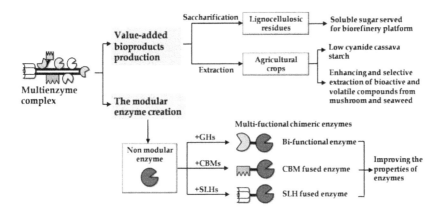

Figure 10. The multienzyme complex of *P. curdlanolyticus* strain B-6 for biotechnological applications.

Biological treatment and saccharification using microorganisms and their enzymes selectively for degradation of lignocellulosic residues has the advantages of low energy consumption, minimal waste production, and environmental friendliness (Schwarz, 2001). The catalytic components of the multienzyme complex release soluble sugars, simple 5- and 6-carbon, from lignocellulose providing the primary carbon substrates, which can be subsequently converted into fuels by microorganisms. For enzyme saccharification, the close proximity between cellulolytic and xylanolytic enzymes is key to concerted degradation of the substrate, whereby the activities of the different enzymes facilitate the activities of their counterparts by promoting access to appropriated portions of the rigid insoluble substrates, since the release of sugar products was high. The synergistic action of the combination of enzymes by different modes of actions (xylanases and cellulases) and the presence of xylan- or cellulose-binding ability on lignocellulose enhanced soluble sugars released from the plant cell walls. In practicality, the multienzyme complex produced from *P. curdlanolyticus* B-6 allows access to lignocellulosic substrate and produces reducing sugar more than non-complexed enzymes from fungi (*T. viride* and *Aspergillus niger*) when the same cellulase activity (0.1 unit) was applied for degradation of corn hull and rice straw residues (unpublished data). In addition, the multienzyme complex of strain B-6 has been used to improve the extraction of plant food such as making low-cyanide-cassava starch by using multienzyme complex to enhance linamarin released by allowing more contact between linamarase and linamarin (Sornyotha et. al., 2009). Also, extraction of volatile compounds such as sea food-like flavor from seaweed, served for food supplement. Consequently, enzymatic treatment has advantages for the preparation of β-glucan and acidic α-glucan-protein complex from the fruiting body of mushroom, *Pleurotus sajor-caju* because the specificity of the multienzyme complex and gentle conditions allow for the recovery of high purity glucans in their native forms with minimal degradation (Satitmanwiwat et al., 2012a,b).

Typically, most plant cell wall degrading enzymes are composed of a series of separate modules (modular enzymes). These domains may fold and function in an independent manner and are normally separated by short linker. *P. curdlanolyticus* B-6, produces a number of glycosyl hydrolase (GHs) families and CBM families which have different substrates recognition affinity and increase amorphous regions of cellulose by H-bond elimination. Interestingly, modular architecture created by chimeric proteins creation with various tandem CBMs, GHs, and SLH-specific, should make it possible to construct effective lignocellulosic degrading enzymes, strongly binding, targeting enzyme to their substrates and bacterial cell surfaces for enhancing a variety of substrates hydrolysis. The strong carbohydrate-binding property of the cellulose-binding domain and xylan-binding domain, specific degradative activities exhibit important properties of the lignocellulosic material degrading enzymes that can be used in biotechnology.

6. Conclusion

A facultatively anaerobic bacterium *P. curdlanolyticus* strain B-6, isolated from an anaerobic digester fed with pineapple wastes, is unique in that it produces extracellular xylanolytic-cellulolytic multienzyme complex capable of efficient degradation of plant biomass materials under aerobic conditions. The production of strain B-6 multienzyme complex under aerobic conditions has several advantages: (i) a simple process, (ii) low price of medium, (iii) high growth rate, (iv) large quantities of extracellular enzymes yields, and (v) safe use with regard to health and environmental aspects. Thus, strain B-6 and its multienzyme complex is a promising tool for an industrial process employing direct hydrolysis for the bioconversion of cellulose as well as hemicellulose in biomass. This review shows that strain B-6 multienzyme complex is a novel enzymatic system known at the biochemical, genetic, and mechanism level. It also stresses that some points still need to be further investigated, mainly (i) the elucidation of scaffolding protein functions, (ii) the characterization of others key enzyme subunits, (iii) the assembly mechanism of the multienzyme complex, (iv) improvement of the efficiency in degradation of biomass of the multienzyme complex, and (v) improvement of the production of the multienzyme complex. The latter will certainly represent a challenge for future research.

Author details

Khanok Ratanakhanokchai, Rattiya Waeonukul, Patthra Pason, Chakrit Tachaapaikoon and Khin Lay Kyu
King Mongkut's University of Technology Thonburi, Thailand

Kazuo Sakka
Mie University, Japan

Akihiko Kosugi and Yutaka Mori
Japan International Research Center for Agricultural Sciences, Japan

7. References

Aro, N.; Pakula, T. & Penttilä, M. (2005). Transcriptional Regulation of Plant Cell Wall Degradation by Filamentous Fungi. *FEMS Microbiology Reviews*, Vol.29, No.4, (September, 2005), pp. 719-739, ISSN 0168-6445

Aspinall, G.O. (1980). Chemistry of Cell-Wall Polysaccharides, In: *The Biochemistry of Plants. A Comprehensive Treatise*, Vol.3, Preiss J., pp. (473–500), Academic Press, ISBN 0-12-675403-9, New York

Bayer, E.A.; Belaich, J.P.; Shoham, Y. & Lamed, R. (2004). The Cellulosomes: Multienzyme Machines for Degradation of Plant Cell Wall Polysaccharides. *Annual Review of Microbiology*, Vol.58, (October, 2004), pp. 521-554, ISSN 0066-4227

Bayer, E.A.; Chanzyt, H.; Lamed, R. & Shoham, Y. (1998). Cellulose, Cellulases and Cellulosomes. *Current Opinion in Structural Biology*, Vol.8, No.5, (October, 1998), pp. 548-557, ISSN 0959-440X

Bayer, E.A. & Lamed, R. (1986). Ultrastructure of the Cell Surface Cellulosome of *Clostridium thermocellum* and Its Interaction with Cellulose. *Journal of Bacteriology*, Vol.167, No.3, (September, 1986), pp. 828-836, ISSN 0021-9193

Bayer, E.A.; Lamed, R. & Himmel, M.E. (2007). The Potential of Cellulases and Cellulosomes for Cellulosic Waste Management. *Current Opinion in Biotechnology*, Vol.18, No.3, (June, 2007), pp. 237-245, ISSN 0958-1669

Bayer, E.A.; Morag, E. & Lamed, R. (1994). The Cellulosome—a Treasure-Trove for Biotechnology. *Trends in Biotechnology*, Vol.12, No.9, (September, 1994), pp. 379-386, ISSN 0167-7799

Bayer, E.A.; Setter, E. & Lamed, R. (1985). Organization and Distribution of the Cellulosome in *Clostridium thermocellum*. *Journal of Bacteriology*, Vol.163, No.2, (August, 1985), pp. 552-559, ISSN 0021-9193

Berger, E.; Jones, W.A.; Jones, D.T. & Woods, D.R. (1990). Sequencing and Expression of a Cellodextrinase (*ced1*) Gene from *Butyrivibrio fibrisolvens* H17c Cloned in *Escherichia coli*. *Molecular and General Genetic*, Vol.223, No.2, (September, 1990), pp. 310-318, ISSN 0026-8925

Bhat, S.; Goodenough, P.W.; Owen, E. & Bhat, M.K. (1993). Cellobiose: A True Inducer of Cellulosome in Different Strains of *Clostridium thermocellum*. *FEMS Microbiology Letters*, Vol.111, No.1, (July, 1993), pp. 73-78, ISSN 0378-1097

Blair, B.G. & Anderson, K.L. (1999a). Regulation of Cellulose Inducible Structures of *Clostridium cellulovorans*. *Canadian Journal of Microbiology*, Vol.45, No.3, (March, 1999), pp. 242-249, ISSN 008-4166

Blair, B.G. & Anderson, K.L. (1999b). Cellulose-Inducible Ultrastructural Protuberances and Cellulose-affinity Proteins of *Eubacterium cellulosolvens*. *Anaerobe*, Vol.5, No.5, (October, 1999) pp. 547-554, ISSN 1075-9964

Borneman, W.S.; Akin, D.E. & Ljungdahl, L.G. (1989). Fermentation Products and Plant Cell Wall Degrading Enzymes Produced by Monocentric and Polycentric Anaerobic Ruminal Fungi. *Applied and Environmental Microbiology*, Vol.55, No.5, (May, 1989), pp. 1066-1073, ISSN 0099-2240

Chimtong, S.; Tachaapaikoon, C.; Pason, P.; Kyu, K.L.; Kosugi, A.; Mori, Y. & Ratanakhanokchai, K. (2011). Isolation and Characterization of Endocellulase-Free Multienzyme Complex from Newly Isolated *Thermoanaerobacterium thermosaccharolyticum* Strain NOI-1. *Journal of Microbiology and Biotechnology*, Vol.21, No.3, (March, 2011), pp. 284-292, ISSN 1017-7825

Dalrymple, B.P.; Cybinski, D.H.; Layton I.; McSweeney,C.; Xue, G.P.; Swadling Y.J. & Lowry, J.B. (1997). Three *Neocallimastix patriciarum* Esterases Associated with the Degradation Complex Polysaccharides are Members of a New Family of Hydrolases. *Microbiology*, Vol.143, No.8, (August, 1997) pp. 2605-2614, ISSN 1350-0872

Dashtban, M.; Schraft, H. & Qin, W. (2009). Fungal Bioconversion of Lignocellulosic Residues; Opportunities & Perspectives. *International Journal of Biological Sciences*, Vol.5, No.6, (September, 2009), pp. 578-595, ISSN 1449-1907

Demain, A.L.; Newcomb, M. & Wu, J.H.D. (2005). Cellulase, *Clostridium*, and Ethanol. *Microbiology and Molecular Biology Reviews*, Vol.69, No.1, (March, 2005), pp. 124–154, ISSN 1092-2172

Ding, S.Y.; Bayer, E.A.; Steiner, D.; Shoham, Y. & Lamed, R. (1999). A Novel Cellulosomal Scaffoldin from *Acetivibrio cellulolyticus* that Contains a Family 9 Glycosyl Hydrolase. *Journal of Bacteriology*, Vol.181, No.21, (November, 1999), pp. 6720-6729, ISSN 0021-9193

Ding, S.Y.; Bayer, E.A.; Steiner, D.; Shoham, Y. & Lamed, R. (2000). A Scaffoldin of the *Bacteroides cellulosolvens* Cellulosome that Contains 11 Type II Cohesins. *Journal of Bacteriology*, Vol.182, No.17, (September, 2000), pp. 4915-4925, ISSN 0021-9193

Ding, S.Y.; Rincon, M.T.; Lamed, R.; Martin, J.C.; Mccrae, S.I.; Aurilia, V.; Shoham Y.; Bayer E.A. & Flint, H.J. (2001). Cellulosomal Scaffoldin-Like Proteins from *Ruminococcus flavefaciens*. *Journal of Bacteriology*, Vol.183, No.6, (March, 2001), pp. 1945–1953, ISSN 0021-9193

Doi, R.H. & Kosugi, A. (2004). Cellulosomes: Plant-Cell-Wall-Degrading Enzyme Complexes. *Nature Reviews Microbiology*, Vol.2, No.7, (July, 2004), pp. 541–551, ISSN 1740-1526.

Doi, R.H.; Kosugi, A.; Murashima, K.; Tamaru, Y. & Han, S.O. (2003). Cellulosomes from Mesophilic Bacteria. *Journal of Bacteriology*, Vol.185, No.20, (October, 2003), pp. 5907–5914, ISSN 0021-9193

Gal, L.; Pages, S.; Gaudin, C.; Belaich, A.; Reverbel-Leroy, C.; Tardif, C. & Belaich, J.P. (1997). Characterization of the Cellulolytic Complex (Cellulosome) Produced by *Clostridium cellulolyticum*. *Applied and Environmental Microbiology*, Vol.63, No.3, (March, 1997), pp. 903–909, ISSN 0099-2240

Goyal, A.; Ghosh, B. & Eveleig, D. (1991). Characterization of Fungal Cellulases. *Bioresource Technology*, Vol.36, No.1, (Special Issue for Enzymatic Hydrolysis of Cellulose, 1991), pp. 37-50, ISSN 0960-8524

Han, S.O.; Cho, H.Y.; Yukawa, H.; Inui, M. & Doi, R.H. (2004). Regulation of Expression of Cellulosomes and Noncellulosomal (Hemi)Cellulolytic Enzymes in *Clostridium cellulovorans* During Growth on Different Carbon Sources. *Journal of Bacteriology*, Vol.186, No.13, (July, 2004), pp. 4218-4227, ISSN 0021-9193

Han, S.O.; Yukawa, H.; Inui, M. & Doi, R.H. (2005). Effect of Carbon Source on the Cellulosomal Subpopulations of *Clostridium cellulovorans*. *Microbiology*, Vol.151, No.5, (May, 2005), pp. 1491-1497, ISSN 1350-0872

Hou, P.; Li, Y.; Wu, B.; Yan, Z.; Yan, B. & Gao, P. (2006). Cellulolytic Complex Exists in Cellulolytic Myxobacterium *Sorangium*. *Enzyme and Microbial Technology*, Vol.38, No.1-2 (January, 2006), pp. 273–278, ISSN 0141-0229

Innis, M.A. & Gelfand, D.H. (1990). Optimization of PCRs, In: *PCR protocols: A Guide to Methods and Application*, Innis, M.A.; Gelfand, D.H.; Sninsky, J.J. & White, T.J. pp. (3-12), Academic Press, ISBN 0-12-372180-6, San Diego

Jiang, Z.Q.; Deng, W.; Li, L.T.; Ding, C. H.; Kusakabe, I. & Tan, S.S. (2004). A Novel, Ultra-Large Xylanolytic Complex (Xylanosome) Secreted by *Streptomyces olivaceoviridis*. *Biotechnology Letters*, Vol.26, No.5, (March, 2004), pp. 431-436, ISSN 0141-5492

Kakiuchi, M.; Isui ,A.; Suzuki, K.; Fujino, T.; Fujino, E.; Kimura, T.; Karita, S.; Sakka, K. & Ohmiya, K. (1998). Cloning and DNA Sequencing of the Genes Encoding *Clostridium josui* Scaffolding Protein CipA and Cellulase CelD and Identification of Their Gene

Products as Major Components of the Cellulosome. *Journal of Bacteriology*, Vol.180, No.16, (August, 1998), pp. 4303–4308, ISSN 0021-9193

Kim, C.H. & Kim, D.S. (1993). Extracellular Cellulolytic Enzymes of *Bacillus circulans* Are Present as Two Multi-Protein Complexes. *Applied Biochemistry and Biotechnology*, Vol.42, No.1, (July, 1993), pp. 83-94, ISSN 0885-4513

Klemm, D.; Heublein, B. & Fink, H.P. (2005). Cellulose: Fascinating Biopolymer and Sustainable Raw Material. *Angewandte Chemie International*, Vol.44, No.22, (May, 2005), pp. 3358-3393, ISSN 1433-7851

Kosugi, A.; Murashima K. & Doi, R.H. (2001). Characterization of Xylanolytic Enzymes in *Clostridium cellulovorans*: Expression of Xylanase Activity Dependent on Growth Substrates. *Journal of Bacteriology*, Vol.183, No.11, (June, 2001), pp. 7037-7043, ISSN 0021-9193

Lamed, R. & Bayer, E.A. (1988). The Cellulosome of *Clostridium thermocellum*, In: *Advances in Applied Microbiology*, Vol.33, Laskin, A.I., pp. (1–46), Academic Press, ISBN 0-12-002633-3, San Diego

Lamed, R.; Naimark. J.; Morgenstern, E. & Bayer, E.A. (1987). Specialized Cell Surface Structure in Cellulolytic Bacteria. *Journal of Bacteriology*, Vol.169, No.8, (August, 1987), pp. 3792-3800, ISSN 0021-9193

Lamed, R.; Setter, E. & Bayer, E.A. (1983). Characterization of a Cellulose-Binding Cellulase-Containing Complex in *Clostridium thermocellum*. *Journal of Bacteriology*, Vol.156, No.2, (November, 1983), pp. 828-836, ISSN 0021-9193

Li, X.L.; Chen, H. & Ljungdahl, L.G. (1997). Two Cellulases, CelA and CelC, from the Polycentric Anaerobic Fungus *Orpinomyces* Strain PC-2 Contain N-terminal Docking Domains for a Cellulase-Hemicellulase Complex. *Applied and Environmental Microbiology*, Vol.63, No.12, (December, 1997), pp. 4721-4728, ISSN 0099-2240

Li, K.; Azadi, P.; Collins, R.; Tolan, J.; Kim, J.S. & Eriksson, K.E.L. (2000). Relationship Between Activities of Xylanases and Xylan Structures. *Enzyme and Microbial Technology*, Vol.27, No.1-2, (July, 2000), pp. 89-94, ISSN 0141-0229

Lynd, L.R.; Weimer, P.J.; van Zyl, W.H. & Pretorius, I.S. (2002). Microbial Cellulose Utilization: Fundamentals and Biotechnology. *Microbiology and Molecular Biology Reviews*, Vol.66, No.3, (September, 2002), pp. 506-577, ISSN 1092-2172

Mayer, F.; Coughlan, M. P.; Mori, Y., & Ljungdahl, L.G. (1987). Macromolecular Organization of the Cellulolytic Enzyme Complex of *Clostridium thermocellum* as Revealed by Electron Microscopy. *Applied and Environmental Microbiology*, Vol.53, No.12, (December, 1987), pp. 2785-2792, ISSN 0099-2240

Mohand-Oussaid, O.; Payot, S.; Guedon, E.; Gelhaye, E.; Youyou, A. & Petitdemange, H. (1999). The Extracellular Xylan Degradative System in *Clostridium cellulolitycum* Cultivated on Xylan: Evidence for Cell-Free Cellulosome Production. *Journal of Bacteriology*, Vol.181, No.13, (July, 1999), pp. 4035-4040, ISSN 0021-9193

Nochur, S.V.; Demain, A.L. & Roberts, M.F. (1992). Carbohydrate Utilization by *Clostridium thermocellum*: Importance of Internal pH in Regulating Growth. *Enzyme and Microbial Technology*, Vol.14, No.5, (May, 1992), pp. 338-349, ISSN 0141-0229

Nochur, S.V.; Roberts, M.F. & Demain, A.L. (1993). True Cellulase Production by *Clostridium thermocellum* Grown on Different Carbon Sources. *Biotechnology Letter*, Vol.15, No.6, (June, 1993), pp. 641-646, ISSN 0141-5492

Ohara. H.; Karita, S.; Kimura, T.; Sakka, K. & Ohmiya, K. (2000). Characterization of the Cellulolytic Complex (Cellulosome) from *Ruminococcus albus*. *Bioscience, Biotechnology, and Biochemistry*, Vol.64, No.2, (February, 2000), pp. 254–260, ISSN 0916-8451

Ohtsukl, T., Suyanto, S., Yazaki, S.U. & Mimura, A. (2005). Production of Large Multienzyme Complex by Aerobic Thermophilic Fungus *Chaetomium* sp. nov. MS-017 Grown on Palm Oil Mill Fibre. *Letters in Applied Microbiology*, Vol.40, No.2, (February, 2005), pp. 111-116, ISSN 0266-8254

Pagès, S.; Bélaïch, A.; Bélaïch, J.P.; Morag, E.; Lamed, R.; Shoham, Y. & Bayer E.A. (1997). Species-Specificity of the Cohesin–Dockerin Interaction Between *Clostridium thermocellum* and *Clostridium cellulolyticum*: Prediction of Specificity Determinants of the Dockerin Domain. *Proteins: Structure, Function, and Bioinformatics*, Vol.29, No.4, (December, 1997), pp. 517-527, ISSN 0887-3585

Pason, P.; Chon, G.H.; Ratanakhanokchai, K.; Kyu K.L.; Jhee, O.H.; Kang, J.; Kim, W.H.; Choi, K.M.; Park, G.S.; Lee, J.S., Park, H.; Rho, M.S. & Lee, Y.S. (2006a). Selection of Multienzyme Complex Producing Bacteria Under Aerobic Cultivation. *Journal of Microbiology and Biotechnology*, Vol.16, No.8, (August, 2006), pp. 1269-1275, ISSN 1017-7825

Pason, P.; Kosugi, A.; Waeonukul, R.; Tachaapaikoon, C.; Ratanakhanokchai, K.; Arai, T.; Murata, Y.; Nakajima, J. & Mori, Y. (2010). Purification and Characterization of a Multienzyme Complex Produced by *Paenibacillus curdlanolyticus* B-6. *Applied Microbiology and Biotechnology*, Vol.85, No.3, (January, 2010), pp. 573-580, ISSN 0175-7598

Pason, P; Kyu, K.L. & Ratanakhanokchai, K. (2006b). *Paenibacillus curdlanolyticus* Strain B-6 Xylanolytic-Cellulolytic Enzyme System That Degrades Insoluble Polysaccharides. *Applied and Environmental Microbiology*, Vol.72, No.4, (April, 2006), pp. 2483-2490, ISSN 0099-2240

Pauly, M. & Keegstra, K. (2008) Cell Wall Carbohydrates and Their Modification as a Resource for Biofuels. *The Plant Journal*, Vol.54, No.4, (May, 2008), pp. 559–568, ISSN 0960-7412

Pérez, J.; Muñoz-Dorado, J.; De-la-Rubia, T. & Martínez, J. (2002). Biodegradation and Biological Treatments of Cellulose, Hemicellulose and Lignin: an Overview. *International Microbiology*, Vol.5, No.2, (June, 2002), pp. 53-63, ISSN 1139-6709

Phitsuwan, P.; Tachaapaikoon, C.; Kosugi, A.; Mori, Y.; Kyu, K.L. & Ratanakhanokchai, K. (2010). A Cellulolytic and Xylanolytic Enzyme Complex from an Alkalothermoanaerobacterium, *Tepidimicrobium xylanilyticum* BT14. *Journal of Microbiology and Biotechnology*, Vol.20, No.5, (May, 2010), pp. 893-903 ISSN 1017-7825

Pohlschröder, M.D.; Leschine, S.B., & Canale-Parola, E. (1994). Multicomplex Cellulase–Xylanase System of *Clostridium papyrosolvens* C7. *Journal of Bacteriology*, Vol.176, No.1, (January, 1994), pp. 70-76, ISSN 0021-9193

Ponpium, P.; Ratanakhanokchai, K. & Kyu, K.L. (2000). Isolation and Properties of a Cellulosome-Type Multienzyme Complex of the Thermophilic *Bacteroides* sp. Strain P-1.

Enzyme and Microbial Technology, Vol.26, No.5-6, (March, 2000), pp. 459-465, ISSN 0141-0229

Qiu, X.; Selinger, B.; Yanke, L.J. & Cheng, K.J. (2000). Isolation and Analysis of Two Cellulase cDNAs from *Orpinomyces joyonii*. *Gene*, Vol.245, No.1, (March, 2000), pp. 119-126, ISSN 0378-1119

Rabinovich, M.L.; Melnik, M.S. & Bolobova, A.V. (2002). Microbial Cellulases (Review). *Applied Biochemistry and Microbiology*, Vol.38, No.4, (July, 2002), pp. 305-322, ISSN 0003-6838

Sabathé, F.; Bélaïch, A. & Soucaille, P. (2002) Characterization of the Cellulolytic Complex (Cellulosome) of *Clostridium acetobutylicum*. *FEMS Microbiology Letters*, Vol.217, No.1 (November, 2002), pp. 15-22, ISSN 0378-1097

Saha, B.C.; (2003). Hemicellulose Bioconversion. *Journal of Industrial Microbiology and Biotechnology*, Vol.30, No.5, (May, 2003), pp. 279-291, ISSN 1367-5435

Sakka, M.; Higashi, Y.; Kimura, T.; Ratanakhanokchai, K. & Sakka K. (2011). Characterization of *Paenibacillus curdlanolyticus* B-6 Xyn10D, a Xylanase That Contains a Family 3 Carbohydrate-Binding Module. *Applied and Environmental Microbiology*, Vol.77, No.12, (June, 2011), pp. 4260-4263, ISSN 0099-2240

Sánchez, C. (2009). Lignocellulosic Residues: Biodegradation and Bioconversion by Fungi. *Biotechnology Advances*, Vol.27, No.2, (March-April, 2009), pp. 185–194, ISSN 0734-9750

Satitmanwiwat, S.; Ratanakhanokchai, K.; Laohakunjit, N.; Chao, L.K.; Chen, S-T.; Pason, P.; Tachaapaikoon, C. & Kyu, K.L. (2012a) Improved Purity and Immunostimulatory Activity of β-(1→3)(1→6)-Glucan from *Pleurotus sajor-caju* Using Cell Wall-Degrading Enzymes. *Journal of Agricultural and Food Chemistry*, Vol.60, Issue 21, (May, 2012), pp. 5423-5430, ISSN 0021-8561

Satitmanwiwat, S.; Ratanakhanokchai, K.; Laohakunjit, N.; Pason, P.; Tachaapaikoon, C. & Kyu, K.L. (2012b) Purification and Partial Characterization of an Acidic α-Glucan-Protein Complex from the Fruiting Body of *Pleurotus sajor-caju* and Its Effect on Macrophage Activation. *Bioscience, Biotechnology, and Biochemitry*, in press, ISSN 0916-8451

Scheller, H.V. & Ulvskov, P. (2010). Hemicelluloses. *Annual Review of Plant Biology*, Vol.61, (June, 2010), pp. 263-289, ISSN 1543-5008

Schwarz, W.H. (2001). The Cellulosome and Cellulose Degradation by Anaerobic Bacteria. *Applied Microbiology and Biotechnology*, Vol.56, No.5-6, (September, 2001), pp. 634-649, ISSN 0175-7598

Shida, O.; Takagi, H.; Kadowaki, K.; Nakamura L.K. & Komagata, K. (1997). Transfer of *Bacillus alginolyticus*, *Bacillus chondroitinus*, *Bacillus curdlanolyticus*, *Bacillus glucanolyticus*, *Bacillus kobensis*, and *Bacillus thiaminolyticus* to the Genus *Paenibacillus* and Emended Description of the Genus *Paenibacillus*. *International Journal of Systematic and Evolutionary Microbiology*, Vol.47, No.2, (April, 1997), pp. 289-298, ISSN 1466-5026

Shoham, Y.; Lamed, R. & Bayer, E.A. (1999). The Cellulosome Concept as an Efficient Microbial Strategy for the Degradation of Insoluble Polysaccharides. *Trends in Microbiology*, Vol.7, No.7, (July, 1999), pp. 275-281, ISSN 0966-842X

Shoseyov, O. & Doi, R.H. (1990). Essential 170-kDa Subunit for Degradation of Crystalline Cellulose by *Clostridium cellulovorans* Cellulase. *Proceedings of the National Academy of Sciences of the United States of America*, Vol.87, No.6, (March, 1990), pp. 2192-2195, ISSN 0027-8424

Shoseyov, O.; Shani, Z. & Levy, I. (2006). Carbohydrate Binding Modules: Biochemical Properties and Novel Applications. *Microbiology and Molecular Biology Reviews*, Vol.70, No.2, (June, 2006), pp. 283-295, ISSN 1092-2172

Sleat, R.; Mah, R. A. & Robinson, R. (1984). Isolation and Characterization of an Anaerobic, Cellulolytic Bacterium, *Clostridium cellulovorans* sp. nov. *Applied and Environmental Microbiology*, Vol.48, No.1, (July, 1984), pp. 88-93, ISSN 0099-2240

Sneath, P.H.A. (1986). Endospore-Forming Gram-Positive Rods and Cocci, In: *Bergey's Manual of Systematic Bacteriology*, Vol. 2, Sneath, P.H.A.; Mair, N.S.; Sharpe, M.E. & Holt, J. G., pp. (1104– 1207), William & Wilkins, ISBN 0-68307-893-3, Baltimore

Sornyotha, S.; Kyu K.L. & Ratanakhanokchai K. (2010). An Efficient Treatment for Detoxification Process of Cassava Starch by Plant Cell Wall-Degrading Enzymes. *Journal of Bioscience and Bioengineering*. Vol.109, No.1, (January, 2010), pp. 9-14, ISSN 1389-1723

Sudo, M.; Sakka, M; Kimura,T.; Ratanakhanokchai, K. & Sakka, K. (2010). Characterization of *Paenibacillus curdlanolyticus* Intracellular Xylanase Xyn10B Encoded by the *xyn10B* gene. *Bioscience, Biotechnology, and Biochemistry*, Vol.74, No.11, (November, 2010), pp. 2358-2360, ISSN 0916-8451

Teunissen, M.J.; Op den Camp, H.J.M.; Orpin, C.G.; Huis in't Veld, J.H.J. & Vogels, G.D. (1991). Comparison of Growth Characteristics of Anaerobic Fungi Isolated from Ruminant and Non-Ruminant Herbivores During Cultivation in a Defined Medium. *Journal of General Microbiology*, Vol.137, No.6, (June, 1991), pp. 1401-1408, ISSN 0022-1287

Thomson, J.A. (1993). Molecular Biology of Xylan Degradation. *FEMS Microbiology Letters*, Vol.10, No.1-2, (January, 1993), pp.65–82, ISSN 0378-1097

Tomme, P.; Warren, R.A.J. & Gilkes; N.R. (1995). Cellulose Hydrolysis by Bacteria and Fungi, In: *Advances in Microbial Physiology*, Vol.37, Poole, R.K., pp. (1-81), Academic Press, ISBN 0-12-027737-9, London

van Dyk J.S.; Sakka, M.; Sakka, K. & Pletschke, B.I. (2009). The Cellulolytic and Hemi-Cellulolytic System of *Bacillus licheniformis* SVD1 and the Evidence for Production of a Large Multi-Enzyme Complex. *Enzyme and Microbial Technology*, Vol.45, No.5, (November, 2009), pp. 372-378, ISSN 0141-0229

Waeonukul, R.; Kyu, K.L.; Sakka, K. & Ratanakhanokchai, K. (2008). Effect of Carbon Sources on the Induction of Xylanolytic-Cellulolytic Multienzyme Complexes in *Paenibacillus curdlanolyticus* Strain B-6. *Bioscience, Biotechnology, and Biochemistry*, Vol.72, No.2, (February, 2008), pp. 321-328, ISSN 0916-8451

Waeonukul, R.; Pason, P.; Kyu, K.L.; Sakka, K.; Kosugi, A.; Mori, Y, & Ratanakhanokchai, K. (2009a). Cloning, Sequencing, and Expression of the Gene Encoding a Multidomain Endo-β-1,4-Xylanase from *Paenibacillus curdlanolyticus* B-6, and Characterization of the Recombinant Enzyme. *Journal of Microbiology and Biotechnology*, Vol.19, No.3, (March, 2009), pp. 277-285, ISSN 1017-7825

Waeonukul, R.; Kyu, K.L.; Sakka, K. & Ratanakhanokchai, K. (2009b). Isolation and Characterization of a Multienzyme Complex (Cellulosome) of the *Paenibacillus curdlanolyticus* B-6 Grown on Avicel Under Aerobic Conditions. *Journal of Bioscience and Bioengineering*, Vol.107, No.6, (June, 2009), pp. 610-614, ISSN 1389-1723

Watthanalamloet, A.; Tachaapaikoon, C.; Lee, Y. S.; Kosugi, A.; Mori, Y.; Tanasupawat, S.; Kyu, K.L. & Ratanakhanokchai, K. (2012). *Amorocellulobacter alkalithermophilum* gen. nov., sp. nov. an Anaerobic Alkalithermophile, Cellulolytic-Xylanolytic Bacterium Isolated from Soil in a Brackish Area of a Coconut Garden. *International Journal of Systematic and Evolutionary Microbiology*, doi: 10.1099/ijs.0.027854-0

How Soil Nutrient Availability Influences Plant Biomass and How Biomass Stimulation Alleviates Heavy Metal Toxicity in Soils: The Cases of Nutrient Use Efficient Genotypes and Phytoremediators, Respectively

Theocharis Chatzistathis and Ioannis Therios

Additional information is available at the end of the chapter

1. Introduction

There are many factors influencing plant biomass, such as soil humidity, soil and air temperature, photoperiod, solar radiation, precipitations, genotype e.t.c. One of the most important factors influencing biomass is soil nutrient availability. Both nutrient deficiency and toxicity negatively affect total biomass and fruit production [1-10]. So, by controlling the optimum levels of nutrient availability in soil, the production of biomass and, of course, the economic benefit (fruit production) for the farmers can be maximized. In the cases of limited nutrient availability in soils, fertilization seems to be the most usual practice adopted by the farmers in order to ameliorate the low nutrient status. However, since: i) during the last two decades the prices of fertilizers have been dramatically increased, and ii) soil degradation and pollution, as well as underground water pollution, are serious consequences provoked by the exaggerate use of fertilizers, a global concern to reduce the use of fertilizers has been developed. So, the best (most economic and ecological) way in our days to achieve maximum yields is by selecting and growing nutrient efficient genotypes, i.e. genotypes which are able to produce high yields (biomass) in soils with limited nutrient availability. Many researchers studied the influence of genotype on biomass and plant growth (nutrient use efficient genotypes) and found impressive results. According to Chapin and Van Cleve (1991) [11], nutrient use efficiency is defined as the amount of biomass produced per unit of nutrient. So, nutrient use efficient genotypes are those having the ability to produce biomass sufficiently under limited nutrient availability. In our research with different olive cultivars, grown under hydroponics, or in soil substrate, we found significant differences concerning macro- and micronutrient utilization efficiency among genotypes [12-13]. Possible reasons for differential nutrient utilization efficiency among genotypes may be: i) the genetic material used, i.e. cultivar (differential nutrient uptake, accumulation and distribution among tissues, mechanisms of cultivars/genotypes), ii) differential colonization of their root system mycorrhiza fungi. Chatzistathis et al. (2011) [14] refer that the statistically significant differences in Mn, Fe and Zn utilization efficiency among three Greek olive cultivars ('Chondrolia Chalkidikis', 'Koroneiki' and 'Kothreiki') may be probably ascribed to the differential colonization of their root system by arbuscular mycorrhiza fungus (AMF) (the percentage root colonization by AMF varied from 45% to 73%).

Heavy metal (Cu, Zn, Ni, Pb, Mn, Cr, Cd) toxicity is a very serious problem in soils suffering from: i) industrial and mine activities [15], ii) the exaggerate use of fertilizers, fungicides and insecticides, iii) acidity, iv) waterlogging, v) other urban activities, such as municipal sewage sludges, vi) the use of lead in petrols, paints and other materials [16]. Under these conditions, plant growth and biomass are negatively affected [17-20]. According to Caldelas et al. (2012) [19], not only growth inhibition happened, but also root to shoot dry matter partitioning (R/S) modified (increased 80%) at Cr toxic conditions in Iris pseudacorus L. plants. Some plant species, which may tolerate very high metal concentrations in their tissues, can be used as hyper-accumulators and are very suitable in reducing heavy metal concentrations in contaminated soils [21]. These species are able to accumulate much more metal in their shoots, than in their roots, without suffering from metal toxicity [22]. By successive harvests of the aerial parts of the hyper-accumulator species, the heavy metals concentration can be reduced [23]. Phytoremediation is an emerging technology and is considered for remediation of inorganic- and organic-contaminated sites because of its cost-effectiveness, aesthetic advantages, and long-term applicability. This technique involves the use of the ability of some plant species to absorb and accumulate high concentrations of heavy metal ions [17]. Some of these species may be a few ones from Brassicaceae family, such as raya (Brassica campestris L.) [17] and Thlaspi caerulescens [23], or from other families, such as spinach (Spinacia oleracea L.) [17], Sedum plumbizincicola [24], Amaranthus hypochondriacus [25], Eremochloa ophiuroides [26], Iris pseudacorus L. [19], Ricinus communis L., plant of Euphorbiaceae family [18]. Finally, the tree species Genipa Americana L. may be used as one with great ability as phytostabilizer and rhizofilterer of Cr ions, according to Santana et al. (2012) [20]. Basically, there are two different strategies to phytoextract metals from soils: the first approach is the use of metal hyper-accumulator species. The second one is to use fast-growing, high biomass crops that accumulate moderate to high levels of metals in their shoots for metal phytoremediation, such as Poplar (Populus sp.) [27-28], maize (Zea mays), oat (Avena sativa), sunflower (Helianthus annuus) and rice (Oryza sativa L.) [25]. Generally, the more high biomass producing is one plant species, the more efficient is the phytoremediation effect. So, in order to enhance biomass production under metal toxicity conditions, different strategies, such as the application of chemical amendments, may be adopted [21]. Since Fe deficiency symptoms may be appeared under Cu and Zn toxicity conditions in some species of Brassicaceae family used for phytoremediation, a good practice is to utilize Fe foliar sprays in order to enhance biomass, thus the phytoremediation effect [29].

All the above mentioned topics, concerning the influence of nutrient deficiency and metal toxicity on plant biomass, as well as the importance of using nutrient use efficient genotypes and cultivars, are within the aim of the present review. Furthermore, the characteristics that should have the plant species used for phytoremediation (fast-growing, high biomass crops) in heavy metal polluted soils are fully analyzed, and the different strategies that should be adopted in order to enhance plant growth and biomass production under so adverse soil conditions are also discussed under the light of the most important and recent research papers.

2. Agronomic, environmental and genotypic factors influencing plant growth

Plant growth (i.e. biomass production) is influenced by many (agronomic environmental and others, such as genetic) factors. Some of the most important factors that influence biomass production are: i) soil humidity, ii) soil and air temperature, iii) air humidity, iv) photoperiod, v) light intensity, vi) soil fertility, i.e. soil nutrient availability, and vii) genotype, and are fully analyzed below.

2.1. Soil humidity

Soil humidity is a very crucial factor influencing root growth, thus nutrient uptake and total biomass. Many plant species are more sensitive in soil humidity shortage during a particular (crucial) period of their growth. In olive trees, if soil humidity shortage happens early spring, shoot elongation, as well as the formation of flowers and fruits, are negatively influenced. If the shortage happens during summer, shoot thickening, rather than shoot elongation, is influenced. Finally, soil humidity shortage reduces olive tree canopy (in order to reduce the transpiration by leaf surface) and favors root system growth (in order to have the ability to exploit greater soil volume and to search for more soil humidity), so that the ratio canopy/root is significantly reduced [30]. On the other hand, under excess soil humidity conditions (waterlogging), when soil oxygen is limited, the root system may suffer from hypoxia, thus, nutrient uptake is negatively influenced. Under extreme anaerobic soil conditions, the presence of pathogen microorganisms, such as *Phytophthora* sp. may lead to root necrosis. According to Therios (2009) [31], for olive trees the mechanism of tolerance to waterlogging is based on the production of adventitious roots near to the soil surface.

2.2. Soil temperature

Soil temperature influences root growth, thus nutrient and water uptake and, of course, biomass production. Most nutrients are absorbed with energy consumption (energetic uptake), so, low and very high soil temperatures negatively influence root growth and nutrient uptake. Furthermore, low soil temperatures induce a water deficit [32].

2.3. Air temperature

Air temperature directly influences photosynthesis, which is the most important physiological function in plants. The optimum temperature for photosynthesis depends on plant species and also on cultivar for the same species. Usually, the optimum temperature for maximum photosynthetic activity is around 25°C for most vegetative species. When temperature exceeds 35°C photosynthesis is inhibited, thus biomass production may be restrained. High temperatures are associated with a high vapor pressure deficit between leaves and the surrounding air. The same applies to fruit, where high temperatures may cause fruit drop in olive trees [31]. On the other hand, low temperatures act negatively in photosynthesis function and starch is redistributed and is accumulated in organs protected from frost, such as roots. Very low temperatures (<-12°C) damage the leaf canopy, shoot and branches of trees [31].

2.4. Air humidity

Low atmosphere humidity speeds up transpiration by leaf surface. Increase of the rate of transpiration causes reduction of vegetative tissues water content, thus depression in the rate of growth and biomass production.

2.5. Photoperiod

Photoperiod is the duration of light in 24 hours and it is one of the most important factors influencing vegetative growth. Plant species whose vegetative growth is mostly influenced by long day conditions are *Populus robusta*, *Ulmus Americana* and *Aesculus hippocastanus* [30].

2.6. Light intensity

Light, together with CO_2, are the two main factors influencing photosynthetic rate. By increasing light intensity up to an optimum limit the maximum photosynthetic rate, so the greatest biomass production can be achieved.

2.7. Nutrient availability

Limited nutrient availability influences negatively biomass production. Nitrogen deficiency strongly depresses vegetation flush. According to Boussadia et al. (2010) [8], total biomass of two olive cultivars ('Meski' and 'Koroneiki') was strongly reduced (mainly caused by a decrease in leaf dry weight) under severe N deprivation, while in an out-door pot-culture experiment with castor bean plants (*Ricinus communis* L.), conducted by Reddy and Matcha (2010) [9], it was found that among the plant components, leaf dry weight had the greatest decrease; furthermore, root/shoot ratio increased under N deficiency [9]. Phosphorus deficiency caused reduced biomass, photosynthetic activity and nitrogen fixing ability in mungbean (*Vigna aconitifolia*) and mashbean (*Vigna radiata*) [33]. Under P deficiency conditions, genotypic variation in biomass production is evident; according to Pang et al. (2010) [34], who studied in a glasshouse experiment the response of ten perennial herbaceous legume species, found that under low P conditions several legumes produced more biomass than lucerne. Nutrient deficiency may cause physiological and metabolism abnormalities in plants, which may lead to deficiency symptoms. There are two categories of symptoms: i) General symptoms, such as limited growth and inability of reproduction (flowering and fruit setting), caused by the deficiency of many necessary macro- or micro-nutrients, and ii) typical, characteristic, deficiency symptoms, such as chlorosis, i.e. yellowing (due to Fe deficiency). In both cases biomass production is depressed. In the study of Msilini et al. (2009) [10], bicarbonate treated plants of *Arabidopsis thaliana* suffered from Fe deficiency displayed significantly lower biomass, leaf number and leaf surface, as compared to control plants, and showed slight yellowing of their younger leaves. Under limited nutrient availability, arbuscular mycorrhiza fungi (AMF) may favor nutrient uptake and thus enhance biomass production. Hu et al. (2009) [35] refer that AMF inoculation of maize plants was likely more efficient in extremely P-limited soils. Generally, root colonization by AMF influences positively plant growth under N, P, or micronutrient deficiency conditions [36].

2.8. Genotypic factors (root morphology and architecture, genetic growth capacity e.t.c.)

According to Bayuelo-Jimenez et al. (2011) [3], under P deficiency, P-efficient accessions of maize plants (*Zea mays* L.) had greater root to shoot ratio, nodal rooting, nodal root laterals, nodal root hair density and length of nodal root main axis, and first-order laterals. In our experiments, we also found differential root system morphology among three Greek olive cultivars (the root systems of 'Koroneiki' and 'Chondrolia Chalkidikis' were less branched and more lateral, and with less root hair development and density, than that of 'Kothreiki', which was richly-branched and with much greater root hair development and density), something which was probably the main reason for the great genotypic variations in nutrient uptake and growth among the three cultivars (Chatzistathis, unpublished data). Singh et al. (2010) [37] found that great differences existed among 10 multipurpose tree species, grown in a monoculture tree cropping system on the sodic soils of Gangetic alluvium in north India, concerning plant height, diameter e.t.c.

3. Physiological roles of nutrients

The absolutely necessary nutrients for plant growth are the following: N, P, K, S, Ca, Mg (macronutrients), Zn, Cu, B, Mn, Fe, Mo (micronutrients). Without one of these nutrients, plant organism can not grow normally and survive. The physiological roles of these nutrients are described in detail below.

3.1. Macronutrients

Nitrogen: It is a primary component of nucleic acids, proteins, amino acids, purines, pyrimidines and chlorophyll. Nitrogen exerts a significant effect on plant growth, as it reduces biennial bearing and increases the percentage of perfect flowers. In olive trees, lack of N leads to decreased growth, shorter length of annual shoots (<10cm), fewer leaves, reduced flowering and decreased yield [31].

Phosphorus: P is a component of high-energy substances such as ATP, ADP and AMP; it is also important for nucleic acids and phospholipids. Phosphorus affects root growth and maturation of plant tissues and participates in the metabolism of carbohydrates, lipids and proteins [31].

Potassium: K plays a crucial role in carbohydrate metabolism, in the metabolism of N and protein synthesis, in enzyme activities, in the regulation of the opening and closing of stomata, thus to the operation of photosynthesis, in the improvement of fruit quality and disease tolerance, in the activation of the enzymes peptase, catalase, pyruvic kinase e.t.c. [31,38].

Calcium: It is the element that participates in the formation and integrity of cell membranes, in the integrity and semipermeability of the plasmalemma, it increases the activity of many enzymes, it plays a crucial role in cell elongation and division, in the transfer of carbohydrates e.t.c. [31,38].

Magnesium: It is part of chlorophyll molecule, it activates the enzymes of Crebs' cycle and it also plays a role in oil synthesis [38].

Sulphur: Sulphur plays role in the synthesis of some amino-acids, such as cysteine, cystine, methionine, as well as in proteins synthesis. It also activates some proteolytic enzymes, such as papaine, bromeline e.t.c. Finally, it is part of some vitamins' molecule and that of gloutathione [31,38].

3.2. Micronutrients

Iron: Iron plays an important role in chlorophyll synthesis, without being part of its molecule. Furthermore, it participates in the molecule of Fe-proteins catalase, cytochrome a, b, c, hyperoxidase e.t.c. In addition to that, it is found in the enzymes nitric and nitrate reductase, which are responsible for the transformation of NO_3^- into NH_4^+, as well as in nitrogenase, which is the responsible enzyme for the atmospheric N capturing [38].

Manganese: Manganese is activator of the enzymes of carbohydrates metabolism, those of Crebs' cycle, and of some other enzymes, such as cysteine desulphydrase, glutamyl transferase e.t.c. It also plays a key-role in photosystem II of photosynthesis, and particularly in the reactions liberating O_2. Finally, Mn acts as activator of some enzymes catalyzing oxidation and reduction reactions [38].

Zinc: Zn plays crucial role in tryptophane biosynthesis, which is the previous stage from IAA (auxin) synthesis (direct influence of Zn on plant growth and biomass production). IAA concentration is significantly reduced in vegetative tissues suffering from Zn deficiency. In addition to the above, Zn is part of some metal-enzymes [38].

Copper: Cu is activator of some enzymes, as well as it is part of enzymes catalyzing oxidation and reducing reactions, such as oxidase of ascorbic acid, lactase, nitrate and nitric reductase e.t.c. [38].

Boron: B plays role in the transfer of sugars along cell membranes, as well as in RNA and DNA synthesis. It also participates to cell division process, as well as to the pectine synthesis [38].

Molybdenum: It is part of the enzyme nitrogenase (capturing of atmospheric N) and nitric reductase (transformation of NO_3^- to NO_2^-). Mo also participates to the metabolism of ascorbic acid [38].

As it is clear from all the above physiological roles of nutrients, the deficiency of even one of them in the mineral nutrition of higher plants depresses their growth, thus biomass production. So, in order to achieve the maximum biomass production, apart from the optimum conditions of all the other environmental and agronomic factors influencing plant growth (temperature, soil humidity, photoperiod, light intensity), it should always be taken care of maintaining the optimum levels of all the necessary soil nutrients. This is usually achieved with the correct fertilization program of the different crops. For example, fruit trees have high demands in K, since fruit production is a K sink and reduces its levels in plant level. According to Therios (2009) [31], potassium plays an important role in olive nutrition. Thus, fruit trees should be periodically fertilized (usually K fertilizers applied during autumn, or winter, and are incorporated into the soils) with enhanced doses of potassium fertilizers (usually K_2SO_4). Apart from chemical fertilizers, organic amendments can be also applied under limited nutrient conditions in order to enhance plant growth. According to Hu et al. (2009) [35], stem length, shoot and root biomass, as well as crop yield of maize were all greatly increased by the application of organic amendments on a sandy loam soil. Apart from the application of chemical fertilizers, organic amendments e.t.c., another modern method to improve yields and to increase biomass is the irrigation of crops with FFC H_2O, a commercial product currently utilized by the agriculture, fishery and food industries in Japan. In the study of Konkol et al. (2012) [39], radish and shirona plants irrigated with FFC H_2O developed larger average leaf area by 122% and greater dry weight and stem length by 39% and 31%, respectively, compared to the plants irrigated with deionized H_2O. FFC H_2O offers agriculturalists a simple and effective tool for the fortification of irrigation waters with micronutrients [39].

4. Nutrient utilization efficiency (NUE): The case of nutrient use efficient genotypes

World population is expected to increase from 6.0 billion in 1999 to 8.5 billion by 2025. Such an increase in population will intensify pressure on the world's natural resource base (land, water, and air) to achieve higher food production. Increased food production could be achieved by expanding the land area under crops and by increasing yields per unit area through intensive farming. Chemical fertilizers are one of the expensive inputs used by farmers to achieve desired crop yields [40]. However, during the last years, the prices of fertilizers have been considerably increased. Furthermore, soil degradation and pollution, as well as underground water pollution, are serious consequences provoked by the exaggerate use of fertilizers during last decades. These two aspects are responsible for the global concern to reduce the use of fertilizers. The best way to do that is by selecting and growing nutrient use efficient genotypes. According to Khoshgoftarmanesh (2009) [41], cultivation and breeding of micronutrient-efficient genotypes in combination with proper agronomic management practices appear as the most sustainable and cost-effective solution for alleviating food-chain micronutrient deficiency.

Nutrient use efficient genotypes are those having the ability to produce high yields under conditions of limited nutrient availability. According to Chapin and Van Cleve (1991) [11] and Gourley et al. (1994) [42], as nutrient utilization efficiency (NUE) is defined the amount of biomass produced per unit of nutrient absorbed. **Nutrient efficiency ratio (NER)** was suggested by Gerloff and Gabelman (1983) [43] to differentiate genotypes into efficient and inefficient nutrient utilizers, i.e. **NER=(Units of Yields, kgs)/(Unit of elements in tissue, kg)**, while **Agronomic efficiency (AE)** is expressed as the additional amount of economic yield per unit nutrient applied, i.e. **AE=(Yield F, kg-Yield C, kg)/(quantity of nutrient applied, kg)**, where F applies for plants receiving fertilizer and C for plants receiving no fertilizer.

Many researchers found significant differences concerning nutrient utilization efficiency among genotypes (cultivars) of the same plant species [1,12,13,40,44-46] Biomass (shoot and root dry matter production) was used as an indicator in order to assess Zn efficient Chinese maize genotypes, grown for 30 days in a greenhouse pot experiment under Zn limiting conditions [1]. NUE is based on: a) uptake efficiency, b) incorporation efficiency and c) utilization efficiency [40]. The uptake efficiency is the ability of a genotype to absorb nutrients from the soil; however, the great ability to absorb nutrients does not necessarily mean that this genotype is nutrient use efficient. According to Jiang and Ireland (2005) [45], and Jiang (2006) [46], Mn efficient wheat cultivars own this ability to a better internal utilization of Mn, rather than to a higher plant Mn accumulation. We also found in our experiments that, despite the fact that the olive cultivar 'Kothreiki' absorbed and accumulated significantly greater quantity of Mn and Fe in three soil types, compared to 'Koroneiki', the second one was more Mn and Fe-efficient due to its better internal utilization efficiency of Mn and Fe (greater transport of these micronutrients from root to shoots) [12] (Tables 1 and 2). Aziz et al. (2011a) [47] refer that under P deficiency conditions, P content of young leaves in *Brassica* cultivars increased by two folds, indicating remobilization of this nutrient from older leaves and shoot. However, differences in P remobilization among *Brassica* cultivars could not explain the differences in P utilization. Phosphorus efficient wheat genotypes with greater root biomass, higher P uptake potential in shoots and absorption rate of P were generally more tolerant to P deficiency in the growth medium [6]. According to Yang et al. (2011) [48], on average, the K efficient cotton cultivars produced 59% more potential economic yield (dry weight of all reproductive organs) under field conditions even with available soil K at obviously deficient level (60 mg/kg).

The possible causes for the differential nutrient utilization efficiency among genotypes and/or species may be one, or combination of more than one, of the following: a) genetic reasons (genotypic ability to absorb and utilize efficiently, or inefficiently, soil nutrients), b) mycorrhiza colonization of the root system, c) differential root exudation of organic compounds favorizing nutrient uptake, d) different properties of rhizosphere, e) other reasons. According to Cakmak (2002) [49], integration of plant nutrition research with plant genetics and molecular biology is indispensable in developing plant genotypes with high genetic ability to adapt to nutrient deficient and toxic soil conditions and to allocate more micronutrients into edible plant products. According to Aziz et al. (2011b) [50], *Brassica* cultivars with high biomass and high P contents, such as 'Rainbow' and 'Poorbi Raya', at low available P conditions would be used in further screening experiments to improve P efficiency in *Brassica*. More specifically, a number of genes have been isolated and cloned, which are involved in root exudation of nutrient-mobilizing organic compounds [51,52]. Successful attempts have been made in the past 5 years to develop transgenic plants that produce and release large amounts of organic acids, which are considered to be key compounds involved in the adaptive mechanisms used by plants to tolerate P-deficient soil conditions [53-55]. However, differential root exudation ability in nature exists among different plant species. According to Maruyama et al. (2005) [56], who made a comparison of iron availability in leaves of barley and rice, the difference in the Fe acquisition ability between these two species was affected by the differential mugineic acid secretion. Chatzistathis et al. (2009) [12] refer that, maybe, a similar mechanism was responsible for the differential micronutrient uptake and accumulation between the Greek olive cultivars 'Koroneiki' and 'Kothreiki'. According to the same authors, differential reduction of Fe^{3+} to Fe^{2+}, or acidification capacity of root apoplast (which associates with the increase of Fe^{3+}-chelate reductase and H-ATPase activities) among three Greek olive cultivars should not be excluded from possible causes for the significant differences observed concerning Fe uptake [14]. Mycorrhiza root colonization may be another responsible factor for the differential micronutrient utilization efficiency among genotypes. According to Citernesi et al. (1998) [57], arbuscular mycorrhiza fungi (AMF) influenced root morphology of Italian olive cultivars, thus nutrient uptake and accumulation, as well as plant growth. In our study with olive cultivars 'Koroneiki', 'Kothreiki' and 'Chondrolia Chalkidikis', we found significant differences concerning root colonization by AMF (that varied from 45% to 73%), together with great differences in uptake and utilization efficiency of Mn, Fe and Zn among them (particularly, 1.5 to 10.5 times greater amount of Mn, Fe and Zn accumulated by 'Kothreiki', compared to the other two cultivars, but the differences in plant growth parameters between the three cultivars were not impressive; this is why the micronutrient utilization efficiency by 'Kothreiki' was significantly lower, compared to that of the other two ones). Finally, the different properties of rhizosphere among genotypes may be another important factor influencing nutrient uptake and utilization efficiency, and of course biomass production. According to Rengel (2001) [58], who made a review on genotypic differences in micronutrient use efficiency of many crops, micronutrient-efficient genotypes were capable of increasing soil available micronutrient pools through changing the chemical and microbiological properties of the rhizosphere, as well as by growing thinner and longer roots and by having more efficient uptake and transport mechanisms.

Soil	Cultivar	Micronutrient	Root	Stem	Leaves
Marl		Mn			
	Kor		50.2b	38.0a	11.8a
	Koth		74.1a	12.8b	13.1a
Gneiss schist					
	Kor		56.5b	34.2a	9.3a
	Koth		81.3a	10.8b	7.9a
Peridotite					
	Kor		44.0b	44.0a	12.0a
	Koth		76.0a	12.9b	11.1a
Marl		Fe			
	Kor		93.7a	3.9a	2.4a
	Koth		98.0a	0.9b	1.1b
Gneiss schist					
	Kor		94.0a	3.7a	2.3a
	Koth		98.8a	0.6b	0.6b
Peridotite					
	Kor		90.8a	7.1a	2.1a
	Koth		98.3a	0.8b	0.9b
Marl		Zn			
	Kor		49.3b	29.6a	21.1a
	Koth		64.4a	15.6b	20.0a
Gneiss schist					
	Kor		59.1b	26.7a	14.2a
	Koth		73.7a	14.3b	12.0a
Peridotite					
	Kor		37.3b	33.9a	28.8a
	Koth		65.3a	18.0b	16.7b

The different letters in the same column symbolize statistically significant differences between the two olive cultivars in each of the three soils, for $P \leq 0.05$ (n=6) (SPSS; t-test).

Table 1. Distribution (%) of the total per plant quantity of Mn, Fe and Zn in the three vegetative tissues (root, stem and leaves) of the olive cultivars 'Koroneiki' and 'Kothreiki', when each one was grown in three soils (from parent material Marl, Gneiss schist. and Peridotite) with different physicochemical properties (Chatzistathis et al., 2009).

Soil	Cultivar	MnUE	FeUE	ZnUE
Marl		mg of the total plant d.w./µg of the total per plant quantity of micronutrient		
	Kor	31.85a	1.73a	77.53a
	Koth	18.68b	0.65b	68.08a
Gneiss schist				
	Kor	39.87a	1.84a	51.04a
	Koth	17.94b	0.44b	49.15a
Peridotite				
	Kor	23.33a	1.19a	61.75a
	Koth	18.00a	0.58b	72.88a

The different letters in the same column symbolize statistically significant differences between the two cultivars in each of the three soils, for $P \leq 0.05$ (n=6) (SPSS; t-test).

Table 2. Nutrient utilization efficiency (mg of the total plant d.w. /µg of the total per plant quantity of micronutrient or mg of the total per plant quantity of macronutrient) of the olive cultivars 'Koroneiki' and 'Kothreiki', when each of them was grown in three soils (from parent material Marl, Gneiss schist. and Peridotite) with different physicochemical properties (Chatzistathis et al., 2009).

5. The influence of heavy metal toxicity on biomass production

Soil heavy metal contamination has become an increasing problem worldwide. Among the heavy metals, Cu, Zn, Mn, Cd, Pb, Ni and Cr are considered to be the most common toxicity problems causing increasing concern. Growth inhibition and reduced yield are common responses of horticultural crops to nutrient and heavy metal toxicity [2]. Nevertheless, sometimes less common responses happen under metal toxicity conditions. For example, in the case of Pb it has been suggested that inhibition of root growth is one of the primary effects of Pb toxicity through the inhibition of cell division at the root tip [59]. Significant reductions in plant height, as well as in shoot and root dry weight (varying from 3.3% to 54.5%), as compared with that of the controls, were found for *Typha angustifolia* plants in different Cr treatments [60]. Furthermore, according to Caldelas et al. (2012) [19], not only growth inhibition happened (reached 65% dry weight) under Cr toxicity conditions, but also root/shoot partitioning increased by 80%. Under Cr stress conditions, it was found that root and shoot biomass of *Genipa americana* L. were significantly reduced [20]. The biomass reduction of *Genipa americana* trees is ascribed, according to the same authors, to the decreased net photosynthetic rates and to the limitations in stomatal conductance. The disorganization of chloroplast structure and inhibition of electron transport is a possible explanation for the decreased photosynthetic rates of trees exposed to Cr stress [20]. In contrast to the above, Cd and Pb applications induced slight or even significant increase in plant height and biomass. The fact that Cd and Pb addition enhanced Ca and Fe uptake suggests that these two nutrients may play a role in heavy metal detoxification by *Typha angustifolia* plants; furthermore, increased Zn uptake may also contribute to its hyper Pb tolerance, as recorder in the increased biomass over the control plants [60]. According to the

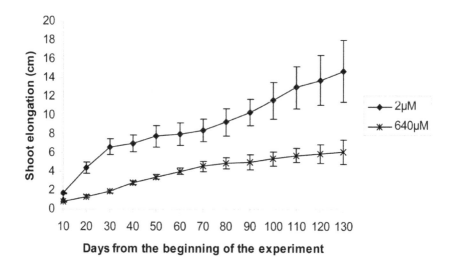

(A)

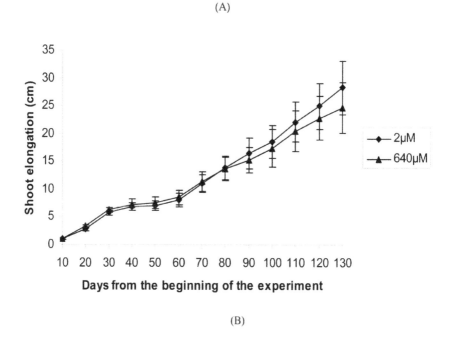

(B)

Figure 1. Shoot elongation of olive cultivars 'Picual' (A) and 'Koroneiki' (B), when grown under hydroponics at normal (2 μM) and excess Mn conditions (640 μM Mn) (Chatzistathis et al., 2012).

same authors (Bah et al., 2011), plants have mechanisms that allow them to tolerate relatively high concentrations of Pb in their environment without suffering from toxic effects.

Tzerakis et al. (2012) [2] found that excessively high concentrations of Mn and Zn in the leaves of cucumber (reached 900 and 450 mg/kg d.w., respectively), grown hydroponically under toxic Mn and Zn conditions, reduced the fruit biomass due to decreases in the number of fruits per plants, as well as in the net assimilation rate, stomatal conductance and transpiration rate. However, it was found that significant differences concerning biomass production between different species of the same genus exist under metal toxicity conditions; *Melilotus officinalis* seems to be more tolerant to Pb than *Melilotus alba* because no differences in shoot or root length, or number of leaves, were found between control plants and those grown under 200 and 1000 mg/kg Pb [15]. In addition to the above, genotypic differences between cultivars of the same species, concerning biomass production, under metal toxicity conditions may also be observed; Chatzistathis et al. (2012) [13] found that under excess Mn conditions (640 µM), plant growth parameters (shoot elongation, as well as fresh and dry weights of leaves, root and stem) of olive cultivar 'Picual' were significantly decreased, compared to those of the control plants (2 µM), something which did not happen in olive cultivar 'Koroneiki' (no significant differences were recorder between the two Mn treatments) (Figure 1). According to the same authors, some factors related to the better tolerance of 'Koroneiki' not only at whole plant level, but also at tissue and cell level, could take place. Such possible factors could be a better compartmentalization of Mn within cells and/or functionality of Mn detoxification systems [13]. Significant growth reductions of several plant species, grown under Mn toxicity conditions, have been mentioned by several researchers [61-65].

Nickel (Ni) toxicity, which may be a serious problem around industrial areas, can also cause biomass reduction. At high soil Ni levels (>200 mg/kg soil) reduced growth symptoms of *Riccinus communis* plants were observed [18]. According to Baccouch et al. (1998) [66], the higher concentrations of Ni have been reported to retard cell division, elongation, differentiation, as well as to affect plant growth and development. Excess Cd, which causes direct or indirect inhibition of physiological processes, such as transpiration, photosynthesis, oxidative stress, cell elongation, N metabolism and mineral nutrition may lead in growth retardation, leaf chlorosis and low biomass production [67]. According to the same authors, Cd stress could induce serious damage in root cells of grey poplar (*Populus x canescens*). Arsenic (As) toxicity may be another (although less common) problem contributing to soil contamination. Repeated and widespread use of arsenical pesticides has significantly contributed to soil As contamination [4]. According to the same authors, plant growth parameters, such as biomass, shoot height, and root length, decreased with increased As concentrations in all soils.

6. Phytoremediation

Soil pollution represents a risk to human health in various ways including contamination of food, grown in polluted soils, as well as contamination of groundwater surface soils [68].

Classical remediation techniques such as soil washing, excavation, and chelate extraction are all labor-intensive and costly [69].

Phytoremediation of heavy metal contaminated soils is defined as the use of living green plants to transport and concentrate metals from the soil into the aboveground shoots, which are harvested with conventional agricultural methods [70]. The technique is suitable for cultivated land with low to moderate metal contaminated level. According to Jadia and Fulekar (2009) [71], phytoremediation is an environmental friendly technology, which may be useful because it can be carried out *in situ* at relatively low cost, with no secondary pollution and with the topsoil remaining intact. Furthermore, it is a cost-effective method, with aesthetic advantages and long term applicability. It is also a safe alternate to conventional soil clean up [17]. However, a major drawback of phytoremediation is that a given species typically remediates a very limited number of pollutants [24]. For example, a soil may be contaminated with a number of potentially toxic elements, together with persistent organic pollutants [72]. There are two different strategies to phytoextract metals from soils. The first approach is the use of metal hyperaccumulator species, whose shoots or leaves may contain rather high levels of metals [25]. The important traits for valuable hyperaccumulators are the high bioconcentration factor (root-to-soil metal concentration) and the high translocation factor (shoot to root metal concentration) [73]. Another strategy is to use fast-growing, high biomass crops that accumulate moderate levels of metals in their shoots for metal phytoremediation [25]. Phytoextraction ability of some fast growing plant species leads to the idea of connecting biomass production with soil remediation of contaminated industrial zones and regions. This biomass will contain significant amount of heavy metals and its energetic utilization has to be considered carefully to minimize negative environmental impacts [74].

7. Plant species used for phytoremediation

Many species have been used (either as hyperaccumulators, or as fast growing-high biomass crops) to accumulate metals, thus for their phytoremediation ability. Hyperaccumulators are these plant species, which are able to tolerate high metal concentrations in soils and to accumulate much more metal in their shoots than in their roots. By successive harvests of the aerial parts of the hyperaccumulator species, the heavy metals concentration in the soil can be reduced [23]. According to Chaney et al. (1997) [21], in order a plant species to serve the phytoextraction purpose, it should have strong capacities of uptake and accumulation of the heavy metals when it occurs in soil solution. For example, *Sedum plumbizincicola* is an hyperaccumulator that has been shown to have a remarkable capacity to extract Zn and Cd from contaminated soils [75]. In addition, a very good also hyperaccumulator for Zn and Cd phytoextraction is *Thlaspi caerulescens* [23]. *Iris pseudacorus* L. is an ornamental macrophyte of great potential for phytoremediation, to tolerate and accumulate Cr and Zn [19]. Furthermore, many species of *Brassica* are suitable for cultivation under Cu and Zn toxicity conditions and may be used for phytoremediation [29]. *Phragmites australis*, which is a species of *Poaceae* family, may tolerate extremely high concentrations of Zn, Cu, Pb and Cd, thus can be used as heavy metal phytoremediator [76].

Santana et al. (2012) [20] refer that *Genipa americana* L. is a tree species that tolerates high levels of Cr^{3+}, therefore it can be used in recomposition of ciliary forests at Cr-polluted watersheds. According to the same authors, this woody species demonstrates a relevant capacity for phytoremediation of Cr. *Elsholtzia splendens* is regarded as a Cu tolerant and accumulating plant species [77]. Peng et al. (2012) [78] refer that *Eucalyptus urophylla X E.grandis* is a fast growing economic species that contributes to habitat restoration of degraded environments, such as the Pb contaminated ones. On the other hand, concerning Cd phytoextraction ability, only a few plant species have been accepted as Cd hyperaccumulators, including *Brassica juncea, Thlaspi caerulescens* and *Solanum nigrum*. Poplar (*Populus* L.), which is an easy to propagate and establish species and it has also the advantages of rapid growth, high biomass production, as well as the ability to accumulate high heavy metal concentrations, could be used as a Cd-hypaeraccumulator for phytoremediation [27-28,67]. According to Wang et al. (2012) [28], the increase in total Cd uptake by poplar genotypes in Cd contaminated soils is the result of enhanced biomass production under elevated CO_2 conditions. Furthermore, *Amaranthus hypochondriacus* is a high biomass, fast growing and easily cultivated potential Cd hyperaccumulator [25]. Another species was found to be a good phytoremediator concerning its phytoaccumulation and tolerance to Ni stress is *Riccinus communis* L. [18]. Finally, *Justicia gendarussa*, which was proved to be able to tolerate and accumulate high concentration of heavy metals (and especially that of Al), could be used as a potential phytoremediator.

Differences between species, or genotypes of the same species, concerning heavy metal accumulation have been found by many researchers. According to Dheri et al. (2007) [17], the overall mean uptake of Cr in shoot was almost four times and in root was about two times greater in rays, compared to fenugreek. These findings, according to the same authors, indicated that family *Cruciferae* (raya) was most tolerant to Cr toxicity, followed by *Chenopodiaceae* (spinach) and *Leguminosae* (fenugreek). Peng et al. (2012) [78] found that cultivar ST-9 of *Eucalyptus urophylla X E.grandis* was shown to accumulate more Pb than others of the same species, like ST-2, or ST-29.

8. Different strategies adopted in order to enhance biomass production under heavy metal toxicity conditions

Under elevated CO_2 conditions the photosynthetic rate is enhanced, thus biomass production is positively influenced. According to Wang et al. (2012) [28], the increase in total Cd uptake by poplar (*Populus* sp.) and willow (*Salix* sp.) genotypes due to increased biomass production under elevated CO_2 conditions suggests an alternative way of improving the efficiency of phytoremediation in heavy metal contaminated soils.

The use of fertilizers is another useful practice that should be adopted by the researchers in order to enhance biomass production under extreme heavy metal toxicity conditions. Some Brassica species, which are suitable to be used as phytoremediators, may suffer from Fe or Mn deficiency symptoms under Cu, or Zn toxicity conditions. In that case, leaf Fe and Mn fertilizations should be done in order to increase their biomass production [29], thus their ability to absorb and accumulate great amounts of heavy metals in contaminated soils, i.e. the efficiency of phytoremediation. According to Li et al. (2012) [25], in order to achieve large biomass crops, heavy fertilization has been practiced by farmers. Application of fertilizers not only provides plant nutrients, but may also change the speciation and mobility of heavy metals, thus enhances their uptake. According to Li et al. (2012) [25], NPK fertilization of *Amaranthus hypochondriacus*, a fast growing species grown under Cd toxicity conditions, greatly increased dry biomass by a factor of 2.7-3.8, resulting in a large increment of Cd accumulation. High biomass plants may be beneficed and overcome limitations concerning metal phytoextraction from the application of chemical amendments, including chelators, soil acidifiers, organic acids, ammonium e.t.c. [21]. Mihucz et al. (2012) [79] found that Poplar trees, grown hydroponically under Cd, Ni and Pb stress, increased their heavy metal accumulation by factor 1.6-3.3 when Fe (III) citrate was used.

Mycorrhizal associations may be another factor increasing resistance to heavy metal toxicity, thus reducing the depression of biomass due to toxic conditions. Castillo et al. (2011) [80] found that when *Tagetes erecta* L. colonized by *Glomus intraradices* displayed a higher resistance to Cu toxicity. According to the same authors, *Glomus intraradices* possibly accumulated excess Cu in its vesicles, thereby enhanced Cu tolerance of *Tagetes erecta* L. [80].

Finally, other factors, such as the influence of *Bacillus* sp. on plant growth, in contaminated heavy metal soils, indicate that biomass may be stimulated under so adverse conditions. According to Brunetti et al. (2012) [81], the effect of the amendment with compost and *Bacillus licheniformis* on the growth of three species of *Brassicaceae* family was positive, since it significantly increased their dry matter. Furthermore, the strain of *Bacillus* SLS18 was found to increase the biomass of the species sweet sorghum (*Sorghum bicolor* L.), *Phytolacca acinosa Roxb.*, and *Solanum nigrum* L. when grown under Mn and Cd toxicity conditions [82].

9. Conclusion and perspectives

Biomass production is significantly influenced by many environmental, agronomic and other factors. The most important of them are air and soil temperature, soil humidity, photoperiod, light intensity, genotype, and soil nutrient availability. Soil fertility, i.e. the availability of nutrients in the optimum concentration range, greatly influences biomass production. If nutrient concentrations are out of the optimum limits, i.e. in the cases when nutrient deficiency or toxicity occurs, biomass production is depressed. Under nutrient deficient conditions, the farmers use chemical fertilizers in order to enhance yields and fruit production. However, since the prices of fertilizers have been significantly increased during the last two decades, a very good agronomic practice is the utilization of nutrient use efficient genotypes, i.e. the utilization of genotypes which are able to produce high yields under nutrient limited conditions. Although great scientific progress has been taken place during last years concerning nutrient use efficient genotypes, more research is still needed in order to clarify the physiological, genetic, and other mechanisms involved in each plant species.

On the other hand, in heavy metal contaminated soils, many plant species could be used (either as hyperaccumulators, or as fast growing-high biomass crops) in order to accumulate metals, thus to clean-up soils (phytoremediation). Particularly, the use of fast growing-high biomass species, such as Poplar, having also the ability to accumulate high amounts of heavy metals in their tissues, is highly recommended, as the efficiency of phytoremediation reaches its maximum. Particularly, since a given species typically remediates a very limited number of pollutants (i.e. in the cases when soil pollution caused by different heavy metals, or organic pollutants), it is absolutely necessary to investigate the choice of the best species for phytoremediation for each heavy metal. In addition to that, more research is needed in order to find out more strategies (apart from fertilization, the use of different *Bacillus* sp. strains, CO_2 enrichment under controlled atmospheric conditions e.t.c.) to enhance biomass production under heavy metal toxicity conditions, thus to ameliorate the phytoremediation efficiency.

Author details

Theocharis Chatzistathis* and Ioannis Therios
Laboratory of Pomology, Aristotle University of Thessaloniki, Greece

10. References

[1] Karim MR, Zhang YQ, Tian D, Chen FJ, Zhang FS, Zou CQ (2012) Genotypic differences in zinc efficiency of Chinese maize evaluated in a pot experiment. J. Sci. Food Agric. DOI 10.1002/jsfa.5672.

[2] Tzerakis C, Savvas D, Sigrimis N (2012) Responses of cucumber grown in recirculating nutrient solution to gradual Mn and Zn accumulation in the root zone owing to excessive supply via the irrigating water. J. Plant. Nutr. Soil Sci. 175: 125-134.

[3] Bayuelo-Jimenez JS, Gallardo-Valdez M, Perez-Decelis VA, Magdaleno-Armas L, Ochoa I, Lynch JP (2011) Genotypic variation for root traits of maize (*Zea mays* L.) from the Purhepecha Plateau under contrasting phosphorus availability. Field Crops Research 121: 350-362.

[4] Quazi S, Datta R, Sarkar D (2011) Effects of soil types and forms of arsenical pesticide on rice growth and development. Intern. J. Environ. Sci. Tech. 8: 445-460.

[5] Sandana P, Pinochet D (2011) Ecophysiological determinants of biomass and grain yield of wheat under P deficiency. Field Crops Res. 120: 311-319.

[6] Yaseen M, Malhi SS (2011) Exploitation of genetic variability among wheat genotypes for tolerance to phosphorus deficiency stress. J. Plant Nuutr. 34: 665-699.

[7] Zribi OT, Abdelly C, Debez A (2011) Interactive effects of salinity and phosphorus availability on growth, water relations, nutritional status and photosynthetic activity of barley (*Hordeum vulgare* L.). Plant Biol. DOI : 10.1111/j. 1438-8677.2011.00450.x

---

* Corresponding Author

[8] oussadia O, Steppe K, Zgallai H, El Hadj SB, Braham M, Lemeur M, Van Labeke MC (2010) Effects of nitrogen efficiency on leaf photosynthesis, carbohydrate status and biomass production in two olive cultivars ('Meski' and 'Koroneiki'). Sci. Hortic. 123: 336-342.

[9] Reddy KR, Matcha SK (2010) Quantifying nitrogen effects on castor bean (*Ricinus communis* L.) development, growth and photosynthesis. Indus. Crops Prod. 31: 185-191.

[10] Msilini N, Attia H, Bouraoui N, M'rah S, Ksouri R, Lachaal M, Ouerghi Z (2009) Responses of *Arabidopsis thaliana* to bicarbonate-induced iron deficiency. Acta Physiol. Plant. 31: 849-853.

[11] Chapin, FS, Van Cleve, K (1991) Approaches to studying nutrient uptake, use and loss in plants. In: Pearcy RW, Ehleringer JR, Mooney HA, Rundel PW, editors. Plant physiological ecology-Field methods and instrumentation. Chapman and Hall, New York. pp 185-207.

[12] Chatzistathis T, Therios I, Alifragis D (2009) Differential uptake, distribution within tissues, and use efficiency of Manganese, Iron and Zinc by olive cultivars 'Kothreiki' and 'Koroneiki'. HortSci. 44: 1994-1999.

[13] Chatzistathis T, Papadakis I, Therios I, Patakas A, Giannakoula A, Menexes G (2012) Differential response of two olive cultivars to excess manganese. J. Plant Nutr. 35: 784-804.

[14] Chatzistathis T, Therios I (2011) Possible reasons for the differential micronutrient utilization efficiency of olive cultivars. Proc. 4th Intern. Olivebioteq seminar. Chania 31 Oct.-4 Nov. 2011, Crete, Greece. Article in Press.

[15] Fernandez R, Bertrand A, Garcia J, Tames RS, Gonzalez A (2012) Lead accumulation and synthesis of non- protein thiolic peptides in selected clones of *Melilotus alba* and *Melilotus officinalis*. Env. Exp. Bot. 78:18-24.

[16] Shen ZG, Li XD, Chen XM, Wang CC, Chua H (2002) Phytoextraction of Pb from a contaminated soil using high biomass species. J. Environ. Qual. 31: 1893-1900.

[17] Dheri GS, Brar MS, Malhi SS (2007) Comparative phytoremediation of chromium-contaminated soils by fenugreek, spinach and raya. Com. Soil Sci. Plant Anal. 38: 1655-1672.

[18] Adhikari T, Kumar A (2012) Phytoaccumulation and tolerance of *Riccinus communis* L. to nickel. Int. J. Phytorem. 14: 481-492.

[19] Caldelas C, Araus JL, Febrero A, Bort J (2012) Accumulation and toxic effects of chromium and zinc in *Iris pseudacorus* L. Acta Physiol. Plant. DOI 10.1007/s11738-012-0956-4.

[20] Santana KB, De Almeida AAF, Souza VL, Mangabeira PAO, Da Silva DC, Gomes FP, Dutruch L, Loguercio LL (2012) Physiological analyses of *Genipa americana* L. reveals a tree with ability as phytostabilizer and rhizofilterer of chromium ions for phytoremediation of polluted watersheds. Env. Exp. Bot. 80: 35-42.

[21] Chaney RL, Malik M, Li YM, Brown SL, Brewer EP, Angle JS, Baker AJM (1997) Phytoremediation of soil metals. Curr. Opin. Biotech. 8: 279-284.

[22] Baker AJM (1981) Accumulators and excluders-strategies in the response of plants to heavy metals. J. Plant Nutr. 3: 643-654.

[23] Frerot H, Petit C, Lefebvre C, Gruber W, Collin C, Escarre J (2003) Zinc and cadmium accumulation in controlled crosses between metallicolous and nonmetallicolous populations of *Thlaspi caerulescens* (Brassicaceae). New Phytologist 157: 643-648.

[24] Wu L, Li Z, Han C, Liu L, Teng Y, Sun X, Pan C, Huang Y, Luo Y, Christie P (2012) Phytoremediation of soil contaminated with cadmium, copper and polychlorinated biphenyls. Intern. J. Phytorem. 14: 570-584.

[25] Li NY, Fu QL, Zhuang P, Guo B, Zou B, Li ZA (2012) Effect of fertilizers on Cd uptake of Amaranthus hypochondriacus, a High Biomass, Fast Growing and Easily Cultivated Potential Cd Hyperaccumulator. Int. J. Phytorem. 14: 162-173.

[26] Liu Y, Wang K, Xu P, Wang Z (2012) Physiological responses and tolerance threshold to cadmium contamination in *Eremochloa ophiuroides*. Intern. J. Phytorem. 14: 467-480.

[27] Jun R, Ling T (2012) Increase of Cd accumulation in five Poplar (*Populus* L.) with different supply levels of Cd. Int. J. Phytorem. 14: 101-113.

[28] Wang R, Dai S, Tang S, Tian S, Song Z, Deng X, Ding Y, Zou X, Zhao Y, Smith DL (2012) Growth, gas exchange, root morphology and cadmium uptake responses of poplars and willows grown on cadmium-contaminated soil to elevated CO_2. Environ Earth Sci DOI 10.1007/s12665-011-1475-0.

[29] Ebbs SD, Kochian L (1997) Toxicity of zinc and copper to Brassica species: Implications for phytoremediation. J. Env. Qual. 26: 776-781.

[30] Voyiatzis D, Petridou M (1997) Biology of horticultural plants. Publications of the Aristotle University of Thessaloniki, Thessaloniki, Greece. pp 51-65.

[31] Therios I (2009) Olives. Crop Production Science in Horticulture. CAB International. UK. pp 179-209.

[32] Pavel EW, Fereres E (1998) Low soil temperatures induce water deficits in olive (Olea europaea L.) trees. Physiol. Plant. 104: 525-534.

[33] Chaudhary MI, Adu-Gyamfi JJ, Saneoka H, Nguyen T, Suwa R, Kanai S, El-Shemy HA, Lightfoot DA, Fujita K (2008) The effect of phosphorus deficiency on nutrient uptake, nitrogen fixation and photosynthetic rate in mashbean, mungbean and soybean. Acta Physiol. Plant. 30: 537-544.

[34] Pang J, Tibbett M, Denton MD, Lambers H, Siddique KHM, Bolland MDA, Revell CK, Ryan MH (2010) Variation in seedling growth of 11 perennial legumes in response to phosphorus supply. Plant Soil 328: 133-143.

[35] Hu J, Lin X, Wang J, Dai J, Cui X, Chen R, Zhang J (2009) Arbuscular mycorrhizal fungus enhances crop yield and P-uptake of maize (*Zea mays* L.): A field case study on a sandy loam soil as affected by long-term P-deficiency fertilization. Soil Biol Biochem 41: 2460-2465.

[36] Blanke V, Wagner M, Renker C, Lippert H, Michulitz M, Kuhn AJ, Buscot F (2011) Arbuscular mycorrhizas in phosphate-polluted soil: interrelations between root colonization and nitrogen. Plant Soil 343: 379-392.

[37] Singh YP, Singh G, Sharma DK (2010) Biomass and bio-energy production of ten multipurpose tree species planted in sodic soils of indo-gangetic plains. J. Forestry Res. 21: 19-24.

[38] Therios I (1996) Mineral nutrition and fertilizers. Dedousis publications. Thessaloniki, Greece. pp 174-177. In Greek.

[39] Konkol NR, McNamara CJ, Bearce-Lee K, Kunoh H, Mitchell R (2012) Novel method of micronutrient application increases radish (*Raphanus sativus*) and shirona (*Brassica Rapa* var. Pekinensis) biomass. J. Plant Nutr. 35: 471-479.

[40] Baligar VC, Fageria NK, He ZL (2001) Nutrient use efficiency in plants. Com. Soil Sci. Plant Anal. 32: 921-950.

[41] Khoshgoftarmanesh AH, Schulin R, Chaney RL, Daneshbakhsh B, Afyuni M (2009) Micronutrient-efficient genotypes fro crop yield and nutritional quality in sustainable agriculture. A review. Agron. Sustain. Dev. 30: 83-107.

[42] Gourley CJP, Allan DL, Rousselle MP (1994) Plant nutrient efficiency: A comparison of definitions and suggested improvement. Plant Soil 158: 29-37.

[43] Gerloff GC, Gabelman WH (1983) Genetic basis of inorganic plant nutrition. In Lauchli A, Bieleski RL, editors. Inorganic Plant Nutrition. Encyclopedia and Plant Physiology New Series, volume 15B. Springer Verlag, New York., pp 453-480.

[44] Papadakis IE, Dimassi KN, Therios IN (2003) Response of 2 citrus genotypes to six boron concentrations: concentration and distribution of nutrients, total absorption and nutrient use efficiency. Aus. J. Agric. Res. 54: 571-580.

[45] Jiang WZ, Ireland CR (2005) Characterization of manganese use efficiency in UK wheat cultivars grown in a solution culture system and in the field. J. Agric. Sci. 143: 151-160.

[46] Jiang WZ (2006) Mn use efficiency in different wheat cultivars. Env. Exp. Bot. 57: 41-50.

[47] Aziz T, Ahmed I, Farooq M, Maqsood MA, Sabir M (2011a) Variation in phosphorus efficiency among *Brassica* cultivars I: Internal utilization and phosphorus remobilization. J. Plant Nutr. 34: 2006-2017.

[48] Yang F, Wang G, Zhang Z, Eneji AE, Duan L, Li Z, Tian X (2011) Genotypic variations in potassium uptake and utilization in cotton. J. Plant Nutr. 34: 83-97.

[49] Cakmak I (2002) Plant nutrition research: Priorities to meet human needs for food in sustainable ways. Plant Soil 247: 3-24.

[50] Aziz T, Rahmatullah M, Maqsood MA, Sabir M, Kanwal S (2011b) Categorization of *Brassica* cultivars for phosphorus acquisition from phosphate rock on basis of growth and ionic parameters. J. Plant Nutr. 34: 522-533.

[51] Richardson AE, Hadobas PA, Hayes JE (2001) Extracellular secretion of Aspergillus phytase from *Arabidopsis* roots enables plants to obtain phosphorus from phytate. Plant J. 25: 641-649.

[52] Takahashi M, Nakanishi H, Kawasaki S, Nishizawa NK, Mori S (2001) Enhanced tolerance of rice to low iron availability in alkaline soils using barley nicotianamine aminotransferase genes. Nat. Biotecnol. 19: 466-469.

[53] Kochian LV (1995) Cellular mechanisms of aluminium toxicity and resistance in plants. Annu. Rev. Plant Physiol. Plant Mol. Biol. 46: 237–260.

[54] Marschner H (1995) Mineral nutrition of higher plants. Academic Press, London.

[55] Lopez-Bucio J, Martinez de la Vega O, Guevara-Garcia A, Errera-Estrella L (2000) Enhanced phosphorus uptake in transgenic tobacco that overproduce citrate. Nat. Biotechnol. 18: 450-453.

[56] Maruyama T, Higuchi K, Yoshida M, Tadano T (2005) Comparison of iron availability in leaves of barley and rice. Soil Sci. Plant Nutr. 51: 1037-1042.

[57] Citernesi AS, Vitagliano C, Giovannetti M (1998) Plant growth and root system morphology of *Olea europaea* L. rooted cuttings as influenced by arbuscular mycorrhizas. J. Hortic. Sci. Biotech. 73: 647-654.

[58] Rengel Z (2001) Genotypic differences in micronutrient use efficiency in crops. Com. Soil Sci. Plant Anal. 32: 1163 1186.

[59] Yang YY, Yung JY, Song WY, Suh HS, Lee Y (2000) Identification of rice varieties with high tolerance or sensitivity to lead and characterization of the mechanism of tolerance. Plant Physiol. 124: 1019-1026.

[60] Bah AM, Dai H, Zhao J, Sun H, Cao F, Zhang G, Wu F (2011) Effects of cadmium, chromium and lead on growth, metal uptake and antioxidative capacity in *Typha angustifolia*. Biol. Trace Elem. Res. 142: 77-92.

[61] Ohki K (1985) Manganese deficiency and toxicity effects on photosynthesis, chlorophyll and transpiration in wheat. Crop Sci. 25: 187-191.

[62] Foy CD, Farina MPW, Oakes AJ (1998) Iron-Manganese interactions among clones of nilegrass. J. Plant Nutr. 21: 978-1009.

[63] Alam, S, Kamei S, Kawai S (2001) Amelioration of manganese toxicity in barley with iron. J. Plant Nutr. 24: 1421-1433.

[64] Quartin VML, Antunes ML, Muralha MC, Sousa MM, Nunes MA (2001) Mineral imbalance due to manganese excess in triticales. J. Plant Nutr. 24: 175-189.

[65] Sarkar D, Pandey SK, Sud KC, Chanemougasoundharam A (2004) In vitro characterization of manganese toxicity in relation to phosphorus nutrition in potato (*Solanum tuberosum* L.). Plant Sci. 167: 977-986.

[66] Baccouch S, Chaoui A, Ferjani EEI (1998) Nickel toxicity: Effects on growth and metabolism of maize. J. Plant Nutr. 21: 577-588.

[67] Dai HP, Shan C, Wei Y, Liang JG, Yang T, Sa WQ, Wei AZ (2012) Subcellular localization of cadmium in hyperaccumulator *Populus* x *canescens*. Afr. J. Biotech. 11: 3779-3787.

[68] Guney M, Zagury GJ, Dogan N, Onay TT (2010) Exposure assessment and risk characterization from trace elements following soil ingestion by children exposed to playgrounds, parks and picnic areas. J. Hazard. Mater. 182: 656-664.

[69] Wu G, Kang HB, Zhang XY, Shao HB, Chu LY, Ruan CJ (2010) A critical review on the bio-removal of hazardous heavy metals from contaminated soils: issues, progress, eco-environmental concerns and opportunities. J. Hazard. Mater. 174: 1-8.

[70] Kumar NPBA, Dushenkov V, Motto H, Raskin I (1995) Phytoextraction: the use of plants to remove heavy metals from soils. Environ Sci Technol 29: 1232-1238.

[71] Jadia CD, Fulekar MH (2009) Phytoremediation of hevy metals: recent techniques. Afr. J. Biotechnol. 8: 921-928.

[72] Xu L, Luo YM, Teng Y, Zhang XL, Wang JJ, Zhang HB, Li ZG, Liu WX (2009) Soil environment quality and remediation in Yangtze River Delta IV. Soil acidification and heavy metal pollution in farmland soils around used electronic device disassembling sites. Acta Pedol. Sin. 46: 833-839. In Chinese.

[73] Teofilo V, Marianna B, Giuliano M (2010) Field crops for phytoremediation of metal-contaminated land. A review. Environ Chem Lett. 8: 1-17.

[74] Syc M, Pohorely M, Kamenikova P, Habart J, Svoboda K, Puncochar M (2012) Willow trees from heavy metals phytoextraction as energy crops. Biomass and Bioenergy 37: 106-113.

[75] Wu LH, Li N, Luo YM (2008) Phytoextraction of heavy metal contaminated soil by *Sedum plumbizincicola* under different agronomic strategies. Proc. 5th Int. Phytotechnol. Conf. pp. 49-50.

[76] Ye ZH, Baker AJM, Wong MH, Willis AJ (2003) Copper tolerance, uptake, and accumulation by *Phragmites australis*. Chemosphere 50: 795-800.

[77] Tang SR, Wilke BM, Brooks RR (2001) Heavy-metal uptake by metal-tolerant *Elsholtzia haichowensis* and *Commelina communis* from China. Com. Soil Sci. Plant. 32: 895-905.

[78] Peng X, Yang B, Deng D, Dong J, Chen Z (2012) Lead tolerance and accumulation in three cultivars of *Eucalyptus urophylla X E. grandis*: implication for phytoremediation. Environ Earth Sci DOI 10.1007/s12665-012-1595-1.

[79] Mihucz VG, Csog A, Fodor F, Tatar E, Szoboszlai N, Silaghi-Dumitrescu L, Zaray G (2012) Impact of two iron(III) chelators on the iron, cadmium, lead and nickel accumulation in poplar grown under heavy metal stress in hydroponics. J. Plant Physiol. 169: 561-566.

[80] Castillo OS, Dasgupta-Schubert N, Alvarado CJ, Zaragoza EM, Villegas HJ (2011) The effect of the symbiosis between *Tagetes erecta* L. (marigold) and *Glomus intraradices* in the uptake of copper (II) and its implications for phytoremediation. New Biotech 29.

[81] Brunetti G, Farrag K, Soler-Rovina P, Ferrara M, Nigro F, Senesi N (2012) The effect of compost and *Bacillus licheniformis* on the phytoextraction of Cr, Cu, Pb and Zn by three *Brassicaceae* species from contaminated soils in the Apulia region, Southern Italy. Geoderma 170: 322-330.

[82] Luo S, Xu T, Chen L, Chen J, Rao C, Xiao X, Wang Y, Zeng G, Long F, Liu C, Liu Y (2012) Endophyte-assisted promotion of biomass production and metal-uptake of energy crop sweet sorghum by plant-growth-promoting endophyte Bacillus sp. SLS18. Appl. Microbiol. Biotechnol. 93:1745–1753.

Permissions

The contributors of this book come from diverse backgrounds, making this book a truly international effort. This book will bring forth new frontiers with its revolutionizing research information and detailed analysis of the nascent developments around the world.

We would like to thank Miodrag Darko Matovic, for lending his expertise to make the book truly unique. He has played a crucial role in the development of this book. Without his invaluable contribution this book wouldn't have been possible. He has made vital efforts to compile up to date information on the varied aspects of this subject to make this book a valuable addition to the collection of many professionals and students.

This book was conceptualized with the vision of imparting up-to-date information and advanced data in this field. To ensure the same, a matchless editorial board was set up. Every individual on the board went through rigorous rounds of assessment to prove their worth. After which they invested a large part of their time researching and compiling the most relevant data for our readers. Conferences and sessions were held from time to time between the editorial board and the contributing authors to present the data in the most comprehensible form. The editorial team has worked tirelessly to provide valuable and valid information to help people across the globe.

Every chapter published in this book has been scrutinized by our experts. Their significance has been extensively debated. The topics covered herein carry significant findings which will fuel the growth of the discipline. They may even be implemented as practical applications or may be referred to as a beginning point for another development. Chapters in this book were first published by InTech; hereby published with permission under the Creative Commons Attribution License or equivalent.

The editorial board has been involved in producing this book since its inception. They have spent rigorous hours researching and exploring the diverse topics which have resulted in the successful publishing of this book. They have passed on their knowledge of decades through this book. To expedite this challenging task, the publisher supported the team at every step. A small team of assistant editors was also appointed to further simplify the editing procedure and attain best results for the readers.

Our editorial team has been hand-picked from every corner of the world. Their multi-ethnicity adds dynamic inputs to the discussions which result in innovative outcomes. These outcomes are then further discussed with the researchers and contributors who give their valuable feedback and opinion regarding the same. The feedback is then collaborated with the researches and they are edited in a comprehensive manner to aid the understanding of the subject.

Apart from the editorial board, the designing team has also invested a significant amount of their time in understanding the subject and creating the most relevant covers. They scrutinized every image to scout for the most suitable representation of the subject and create an appropriate cover for the book.

The publishing team has been involved in this book since its early stages. They were actively engaged in every process, be it collecting the data, connecting with the contributors or procuring relevant information. The team has been an ardent support to the editorial, designing and production team. Their endless efforts to recruit the best for this project, has resulted in the accomplishment of this book. They are a veteran in the field of academics and their pool of knowledge is as vast as their experience in printing. Their expertise and guidance has proved useful at every step. Their uncompromising quality standards have made this book an exceptional effort. Their encouragement from time to time has been an inspiration for everyone.

The publisher and the editorial board hope that this book will prove to be a valuable piece of knowledge for researchers, students, practitioners and scholars across the globe.

List of Contributors

Alessandra Trinchera, Carlos Mario Rivera, Andrea Marcucci and Elvira Rea
Agricultural Research Council - Research Centre for Plant-Soil System (CRA-RPS)., Rome, Italy

María Gómez-Brandón, Marina Fernández-Delgado Juárez and Heribert Insam
University of Innsbruck, Institute of Microbiology, Innsbruck, Austria

Jorge Domínguez
Departamento de Ecoloxía e Bioloxía Animal, Facultade de Bioloxía, Universidade de Vigo, Vigo, Spain

Małgorzata Makowska, Marcin Spychała and Robert Mazur
Poznan University of Life Sciences, Department of Hydraulic and Sanitary Engineering, Poznań, Poland

Pengkang Jin, Xin Jin, Xianbao Wang, Yongning Feng and Xiaochang C. Wang
School of Environment & Municipal Engineering, Xi'an University of Architecture & Technology, Xi'an, P. R. China

Duminda A. Gunawardena and Sandun D. Fernando
Department of Biological and Agricultural Engineering, Texas A&M University, College Station, Texas, USA

Moses Isabirye
Faculty of Natural Resources and Environment, Namasagali Campus, Busitema University, Kamuli, Uganda

D.V.N Raju
Research and Dev't Section - Agricultural Department Kakira Sugar Limited, Jinja, Uganda

M. Kitutu
National Environment Management Authority, Kampala, Uganda

V. Yemeline
UNEP/GRID-Arendal, Norway

J. Deckers and J. Poesen
Katholieke Universiteit Leuven, Department of Earth and Environmental Sciences, Celestijnenlaan Leuven, Belgium

T. P. Basso, T.O. Basso, C.R. Gallo and L.C. Basso
University of São Paulo, "Luiz de Queiroz" College of Agriculture, Brazil

Martin Rulík, Adam Bednařík, Václav Mach, Lenka Brablcová, Iva Buriánková, Pavlína Badurová and Kristýna Gratzová
Department of Ecology and Environmental Sciences, Laboratory of Aquatic Microbial Ecology, Faculty of Science, Palacky University in Olomouc, Czech Republic

Khanok Ratanakhanokchai, Rattiya Waeonukul, Patthra Pason, Chakrit Tachaapaikoon and Khin Lay Kyu
King Mongkut's University of Technology Thonburi, Thailand

Kazuo Sakka
Mie University, Japan

Akihiko Kosugi and Yutaka Mori
Japan International Research Center for Agricultural Sciences, Japan

Theocharis Chatzistathis and Ioannis Therios
Laboratory of Pomology, Aristotle University of Thessaloniki, Greece

Printed in the USA
CPSIA information can be obtained
at www.ICGtesting.com
JSHW011440221024
72173JS00004B/882